The

Reference

Shelf

Energy Policy

Edited by Martha Hostetter

The Reference Shelf
Volume 74 • Number 2

The H.W. Wilson Company
2002

The Reference Shelf

The books in this series contain reprints of articles, excerpts from books, addresses on current issues, and studies of social trends in the United States and other countries. There are six separately bound numbers in each volume, all of which are usually published in the same calendar year. Numbers one through five are each devoted to a single subject, providing background information and discussion from various points of view and concluding with a subject index and comprehensive bibliography that lists books, pamphlets, and abstracts of additional articles on the subject. The final number of each volume is a collection of recent speeches, and it contains a cumulative speaker index. Books in the series may be purchased individually or on subscription.

Library of Congress has cataloged this serial title as follows:

Energy policy / edited by Martha Hostetter.
      p. cm.—(Reference shelf; v. 74, no. 2)
      Includes bibliographical references and index.
      ISBN 0-8242-1011-5 (pbk.)
      1. Energy policy. 2. Energy policy—United States. I. Hostetter, Martha. II. Series.

HD9502.A2 E548 2002
333.79—dc21

2002016795

Visit H.W. Wilson's Web site: www.hwwilson.com

Printed in the United States of America

Contents

List of Figures

Preface

Demand for energy has skyrocketed over the past decade in the U.S. and abroad, spurred by expanding economies, burgeoning populations, and new technologies. In the U.S., the Internet alone accounts for 10 percent of energy consumption. This trend is not expected to slow down any time soon: the Energy Information Administration predicts growth rates of up to 60 percent over the next 20 years. In addition to rising energy consumption, the past few decades have seen energy shortages, wild price fluctuations and fuel riots, and growing concerns over the long-term effects of energy use, prompting many to question whether the world's current patterns of energy consumption can continue. At the same time, since Western society's progress and economic strength have been built on unchecked energy consumption, many are also asking what would happen if energy habits changed.

Meeting energy demands is the source of most air pollution, acid rain, toxic contamination of ground water, and radioactive wastes. In addition, most scientists now agree that global warming is caused, at least in part, by the buildup of greenhouse gases created through burning fossil fuels—primarily carbon dioxide, methane, and nitrous oxide. The heat-trapping effects of these gases threaten to catastrophically alter the earth's climate.

Such environmental problems are compounded in the developing world, where populations and cities—and thus energy demand—are growing exponentially. Western nations consumed the vast majority of energy during the last century, but within a few decades, the energy demands of the developing countries of Asia, Latin America, and Africa are expected to meet and surpass those of the West. How countries like China and India meet their energy needs will have an enormous impact on the world's energy trade, and on the environment. As part of the 1997 Kyoto Protocol, developed countries agreed to reduce their own greenhouse gas emissions and to help guide less-developed nations toward cleaner, sustainable energy sources. (In 2001, the United States withdrew from the agreement, citing flaws.)

Many observers predict that, during the 21st century, the world's energy culture will undergo a transformation. Thanks to new technologies, fossil fuel production is becoming cleaner and more efficient. Energy sources like solar and wind are emerging as viable alternatives, and nuclear power, once thought to be on its way out, is once again being heralded as an energy solution. The most excitement—and the least controversy—surrounds hydrogen, which Jules Verne imagined over a century ago would be the "coal of the future."

This book examines the policies that shape energy production today and the forces that will shape the energy needs of tomorrow. The first section looks at the three major fossil fuels—coal, oil, and natural gas, which produce the overwhelming majority of the world's energy. The second section explores alternatives to fossil fuels and traditional modes of production, including renewables and micropower—the production of energy in smaller quantities and closer to consumers. The third section considers two such alternatives in detail—nuclear energy and hydrogen energy. The final section looks at how energy plays a role on the world stage.

I wish to thank the authors and publications that granted permission for their work to be reprinted in this volume. I would also like to acknowledge the contributions of all those at H.W. Wilson who helped to produce this book: Lynn Messina, Sandra Watson, Jennifer Peloso, and Rich Stein.

Martha Hostetter
February 2002

I. Fossil Fuels

Energy consumption by fuel, 1970–2020
(quadrillion Btu)

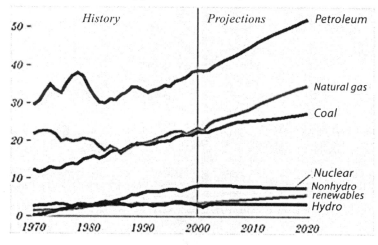

History: Energy Information Administration, *Annual Energy Review 2000*, DOE/EIA-0384(2000) (Washington, DC, August 2001). Projections: Tables A1 and A18

Editor's Introduction

During the twentieth century, in order to travel farther and faster, and to exchange ideas and products across continents, the world's nations consumed 10 times more energy than in any previous century. In particular, they used coal, oil, and natural gas—the "fossil fuels" that were created over millions of years through the decomposition of animal and plant remains. Today, fossil fuels account for between 80 and 90 percent of energy use, worldwide.

While fossil fuels have formed the foundation of modern industrialized society, in recent decades these energy sources have come under fire. Environmentalists label such fuels "non-renewable" to underscore the fact that there are limited amounts of them (and thus we cannot depend on them indefinitely) and to contrast them with energy sources like solar and wind, which are theoretically unlimited. The burning of fossil fuels to run cars, heat homes, and power industries generates a great deal of carbon dioxide, one of the primary greenhouse gases.

Oil is the world's most popular energy source. While nobody disputes the fact that oil reserves are finite, there is widespread disagreement over how much is left, and how quickly the world might run out. In the first article of this section, "Bulls and Bears Duel Over Supply," David Brown outlines the two sides in this debate: Some scientists believe that oil production will peak within a decade, sparking high prices and global instability, while others believe that new exploration tools, together with unconventional sources—including tar sands, oil shale, and heavy oil—will provide modern society with oil for a great many years to come.

In *Popular Science*, David Helvarg reports from the front lines of new oil exploration. Helvarg visits Pompano, an offshore oil rig in the Gulf of Mexico that can drill thousands of feet below the ocean floor, finding oil in areas that were once thought to be tapped out. Over the past five years, deep-sea drilling technology has created a 50 percent spike in domestic oil production in this region, where 93 percent of all offshore drilling takes place.

Coal is the world's second-largest energy source, and unlike oil, it is in abundant supply in America. Today, coal powers more than half of the U.S. electricity grid, and President Bush's proposed energy plan would devote the lion's share of research dollars to new coal exploration and technology—$2 billion over the next 10 years. Advocates claim that new technologies make coal a less-polluting, more efficient energy source. In *Business Week*, Patrick McGuire reports on a Florida power plant that brings together two clean-coal technologies: gasification (turning coal into a cleaner, synthetic gas) and heat recycling, a way of reaping more energy from less coal. Environmentalists contend, however, that such technologies do nothing to limit the amount of carbon dioxide created from burning coal, which is today the single largest contributor to global warming.

Natural gas is less polluting than oil or coal, and thus some see it as a "transition" fuel toward cleaner, renewable energy sources. While today it lags behind oil and coal, it is the fastest-growing energy source worldwide. The newest natural gas plants, which combine combustion and steam, are highly efficient and produce little air pollution. In her report for the *Wall Street Journal*, Ann Zimmerman describes the natural gas boom town of Fairfield, Texas, where rising gas prices and increased demand have created a flourishing economy. Zimmerman's report puts a human face on the debates over fossil fuels by demonstrating how coal, oil, and natural gas are the bread and butter of many communities—one reason so many politicians are eager to defend their continued use.

Bulls and Bears Duel Over Supply[1]

BY DAVID BROWN
AAPG EXPLORER, MAY 2000

The world is running out of crude oil.

No, it isn't.

And there you have the two sides of a simmering controversy in petroleum geology and the oil industry in general.

Do the pessimists have a point?

In his paper "Big Geology for Big Needs," published in the February 1964 AAPG *Bulletin*, A. I. Levorsen projected total United States oil demand of 22 million barrels per day in 2000, with U.S. oil supply of 14 million barrels per day.

Levorsen's demand number appears remarkably accurate, only slightly higher than current forecasts. But his supply number contrasts sharply with average 1999 U.S. oil production of just over 5.9 million barrels per day.

The difference? Levorsen expected a much higher rate of oil discovery during the past 35 years, and anticipated a much lower level of imports.

Supply pessimists take the industry's declining oil discovery rate as an alarm signal, portending a crisis on the near horizon. Most believe world oil production will peak within a decade—if it hasn't already.

"The peak of production will be a more important occurrence than any other event in human history, affecting more people, in more places," said AAPG member Walter Youngquist, a leading proponent of the scarcity theory.

Youngquist is one of a small, persistent group of geologists seeking public and government awareness of this view. L. F. "Buzz" Ivanhoe calls them the Cassandras, after the mythological Trojan princess who could foretell the future but was doomed never to be believed.

"There are a number of us who are petroleum geologists, upstream petroleum professionals. We're all retired, so we can say what we want to," Youngquist said.

These geoscientists offer several points about the outlook for the world's future oil supply:

- The rate of oil discovery is declining, especially for giant and supergiant fields.
- Demand for oil is increasing and will continue to grow.

- Current reserve estimates can't be trusted, since they tend to be exaggerated for political or business reasons.

- Alternate oil sources are inadequate to offset any meaningful reduction in traditional crude supplies.

Cassandras Crossing

Ivanhoe said he began his career as a mining engineer in Ecuador, then returned to the United States and earned a degree in geology at Stanford University. He worked for Chevron (at that time Standard Oil of California), and later Occidental Petroleum.

Today Ivanhoe lives in Ojai, Calif., but serves by long-distance as director of the M. King Hubbert Center for Petroleum Studies at the Colorado School of Mines. The center publishes a quarterly newsletter on the oil-supply outlook, aimed at national legislators and other decision makers.

Pessimists ... foresee a world production peak sometime in the next 10 years.

"At the Hubbert Center we try to get people who know what they're doing, and take their work and reduce it to what I call the 'sports page' level of writing," he said. "For a technical person, that can be hard to do.

"I don't expect any Congressmen to read it, but I hope their aides read it and save it in a file. Then when things change, they'll have all the material on hand."

If the supply pessimists have a guiding spirit, it surely is the ghost of King Hubbert. A widely known and respected Amerada and later Shell Oil geoscientist, Hubbert developed a model for the production lifespan of a finite, nonrenewable resource—the famous "Hubbert curve."

Hubbert first predicted in 1948 that worldwide supplies of oil were limited and that the United States would run short of oil in the 1970s, making the nation dependent on non-U.S. oil. He repeated his warning in 1956, 1962 and 1967.

He later became a research geophysicist with the U.S. Geological Survey and also taught at Stanford University, retiring in 1976.

He died in 1989 at the age of 86.

Hubbert applied this bell-shaped curve to U.S. oil production in 1956 and predicted that domestic supply would begin to decline at the end of the 1960s. Production in the Lower 48 states did peak about 1970 and has declined ever since.

The pessimists take a similar approach to predicting the rise and eventual decline of world oil production. For the most part, they foresee a world production peak sometime in the next 10 years.

"The peak in (world oil) finding was back in the 1960s, and if you go on finding one barrel for every four barrels you produce, pretty soon you're out of business," Ivanhoe said.

"The critical factor for production is the giant fields and the supergiant fields," he added. "They were found many years ago and they are still going strong."

But, he added, we aren't finding them anymore.

Declining Trends

Colin J. Campbell writes and lectures extensively on the oil-supply outlook, and he may be the world's best-known advocate of the declining-production theory.

Campbell said he began his career with Texaco in Trinidad after earning a degree in geology at Oxford University "about 100 years ago."

Actually, he's been an AAPG member since 1959. After joining British Petroleum, and then Amoco, Campbell said he ended his industry career as executive vice president for Fina in Norway. Today he lives in Ballydehob, County Cork, Ireland.

In 1998, Campbell and Jean H. Laherrere, a former Total explorationist, wrote an article titled "The End of Cheap Oil" for *Scientific American* magazine. Some readers thought the article implied the end of oil—a viewpoint Campbell disclaims.

"Many people ask, 'When does the oil run out?' This really isn't the question, because the tail end of depletion goes on forever," Campbell said. "It becomes sort of irrelevant whether you have a few barrels around or not."

In their article, Campbell and Laherrere wrote: "About 80 percent of the oil produced today flows from fields that were found before 1973, and the great majority of them are declining."

"When does the oil run out? . . . The tail end of depletion goes on forever."— **Colin J. Campbell, oil expert**

Youngquist, this year's AAPG's Journalism Award winner, said new exploration tools haven't improved the oil discovery rate, even though today's tools and techniques are much improved from earlier decades.

Even advanced seismic hasn't aided the hunt for giant fields, he added.

"It really isn't like it's making a difference, and I'm well acquainted with 3-D and 4-D seismic," he said. "If you look at 4-D seismic, it even makes it possible to produce oil faster, so you hit a peak sooner."

Supply skeptics doubt the usefulness of forecasts based on reserve totals reported by countries and companies. Those reserve numbers are almost always manipulated and may be wildly inaccurate, they say.

"The former Soviet states are still reporting the same amount of oil reserves they had 10 years ago. China is reporting the same amount as 10 years ago," Ivanhoe said. "It's ridiculous."

Campbell sees a more complex picture, with some countries overestimating reserves while most companies underestimate reserves.

"In the case of the U.K. you have under-reporting, then you go back to OPEC where you had over-reporting back in the 1980s for quota reasons," he observed. "Companies have consistently under-reported their reserves of oil. This is no conspiracy—it's done for sound business reasons."

He believes the United States tends to overstate its oil reserves, and the world's reserves, to downplay the threat of OPEC market control. OPEC countries also tend to overstate their oil holdings for political and economic reasons, he said.

Also, Campbell objects to the practice of adding field-growth numbers to current discoveries, or using them as a basis for supply projections. The U.S. Geological Survey, which received input from the AAPG Committee on Resource Evaluation, announced it will include field-growth estimates in its new "World Petroleum Assessment 2000."

"Of course, it's a very political issue for them," Campbell said of the USGS. "They've taken the U.S. experience and added it not only to the world's reserves today, but also to anything that's discovered in the future. That's absolutely wrong."

Future Shock?

Youngquist said he received a doctorate in geology from the University of Iowa and spent most of his industry career with Exxon, first in Peru and then in the United States. He taught geology at the University of Oregon in Eugene, where he still lives, and later served as a consultant with Exxon, Amoco and Shell.

After studying alternate oil sources, he's concluded they can do little to offset a decline in production of traditional oil.

"Basically, the problem comes when you go from flowing oil or pumping oil to mining the stuff," he said. "To produce any real quantities of oil from these sources just isn't in the cards, I think."

However, Youngquist conceded that oil sands production "will help, especially in Canada, where they're in a pretty good situation." The future rate of production may be dictated by a combination of economic, environmental and engineering factors.

"The issue is primarily the rate of extraction, which is partly economic, with some environmental questions there," Campbell said.

"As of today, only the most favorable deposits have been exploited," he continued, "but the question is, when you go down 75 meters or 150 meters, how economic will it be?"

For the immediate future, Campbell sees OPEC's share of world oil production possibly rising to 35 percent next year. It will ultimately reach 50 percent, at which point OPEC's own production will decline, he predicted.

"At a certain point, and nobody is sure where it will be, this growth in share is translated into control of price," he said.

"Then there's a price shock, which will curb demand."

Youngquist expects governments to begin rationing oil as supply deteriorates.

"What will happen is that price will move oil to its highest and best use," he said. "People ask, 'How long will oil last?' That's inconsequential. When the peak occurs, you have a continuing world oil crisis."

Levorsen's 1964 paper did not attempt to predict U.S. oil supply and demand in 2000. It simply used a projection to show what kind of exploration success the U.S. would need to maintain an oil supply about 50 percent from primary production, 25–30 percent from enhanced recovery and 15–20 percent from imports.

As the title implied, Levorsen thought big oil needs would require Big Geology. In his paper he wrote these words about himself and his fellow petroleum geologists:

"If we cannot discover new oil, competitively, we will go the way of the dinosaurs."

Oil and Water[2]

BY DAVID HELVARG
POPULAR SCIENCE, AUGUST 2001

We fly out of Venice, Louisiana, with its flood-pain levees, low-lying cow pastures, and oil-patch heliports. Our Bell 412 helicopter, a buffed-up version of Huey, quickly passes over shredding islands of brown spartina, or salt hay, cross-hatched with canals and studied with oil tank transfer stations.

We cross the southwest channel of the Mississippi and a surf line the color of chocolate mousse. As we fly beyond the first cluster of oil platforms, the water turns a strange jade green. Soon we're some 50 miles offshore in deep blue water. Everywhere on the horizon are oil platforms. There are 4,000 of these structures in the Gulf of Mexico today.

We circle a flat-topped platform called Pompano. Owned by BP, it's the second tallest bottom-fixed structure in the world, drilling into the floor 1,310 feet below the surface.

While President Bush's talk of an energy crisis and plans for increased production of fossil fuels have sparked controversy, few on either side of the debate have paid much attention to the technologically driven boom in oil and gas production taking place in the deep waters of the Gulf.

Today, offshore drilling accounts for 26 percent of U.S. oil and natural gas production. And despite of the drilling in the Arctic National Wildlife Refuge and in waters off the West Coast, Florida, and North Carolina, 93 percent of offshore production continues to take place in the Gulf. In the early 1990s, there were reports that the Gulf might be a "Dead Sea," tapped out after 50 years of exploitation, but that was before, deep-water drilling technology took off, increasing Gulf oil production by 50 percent in the past five years alone. Today, 52 percent of the oil and 20 percent of the natural gas extracted from the Gulf comes from wells drilled in water depths of 1,000 feet or greater.

Drilling is rapidly moving toward depths of 10,000 feet or more, with Unocal recently sinking a well in 9,743 feet of water. That's too deep for a bottom-fixed platform, but oil companies are extending their reach with new technologies such as cable-stayed and water-filled platforms, and proposals to replace oil rigs with production ships moored to the seafloor.

About 700 feet wide at its base, Pompano is taller than the Empire State Building. We land on the helideck 12 stories above the water. Even with the copter's rotor stopped, the sea winds continue to whip against us at 30 knots. We climb down two levels past some rigid enclosed lifeboats to the living quarters, walking on cookie-cutter grating that lets you see all the way down to the swells breaking against the platform's legs.

Entering the crew structure, we pass a three-button emergency panel marked "Abandon Platform, Fire, and General Quarters." The galley with its cafeteria-style service, metal tables, bug juice dispenser, video player, and thick couches grouped around the oversized TV reminds me of a number of work boats I've been on, minus the sense of ocean movement.

"It's just like an aircraft carrier in that the platform has to be completely self-sufficient," Hugh Depland, BP's public relations guy, tells me.

Normally operated by a crew of 12 (who work seven days on, seven off), Pompano is crowded with 22 extra men who are reconfiguring the platform for the return visit of a drilling rig. After five years of operation, Pompano's production has declined from about 68,000 barrels of oil a day to around 46,000 barrels (and 63 million cubic feet of natural gas). Not bad, at over a million dollars' worth of product every 24 hours, but Pompano can do better, and will. The oil companies are now able to find oil- and gas-laden sands they once missed—using 3-D seismic imaging and computer-controlled bottom sensors. And for older platforms like Pompano, the companies use what they call 4-D seismic studies, incorporating past production patterns into their computer analysis of where additional hydrocarbons might be found.

Down in the MCC, the highly automated Multi-Control Center, I meet George Yount, the operations supervisor. He's wearing a tan Carhartt work coat and BP hard hat and looks like a "beach master" elephant seal, thick throated, well padded but strong, with a scraggly mustache and three day growth of beard. He's been in the industry 25 years.

Also working here is Wendy Lemoine, a thin, blond assistant engineer. While the oil patch has been racially integrated for some time, it's well behind the times when it comes to women. Wendy is the only female among some 80 men on the two platforms I'll visit out here, a fairly typical ratio. A chemical engineer on temporary duty, she says that while she doesn't mind the work, she's definitely looking forward to getting back to Houston where she's based.

After making sure I have a hard hat and ear plugs, George takes me down to the well bay to see the Christmas trees (well pipes). On the way I look over the side and spot about 200 good-size fish schooling around one of the yellow platform legs. A little further out, the torpedo-shaped bodies and yellow tails of a pair of dolphin fish (mahi mahi) streak by. Later in the day we'll spot a big manta ray cruising the area, its 9-foot wings clearing the water like sails. While platforms haven't been shown to increase fish productivity, they do tend to concentrate fish, as do any structures in the ocean, be they coral reefs, shipwrecks, or simply barrels of waste.

The Christmas trees are 23 vertical well pipes (plus two water reinjection pipes) married to small chokes and connectors so the oil can be separated (through "heater/treater" processors) and the gas de-watered before being pumped into big 12-inch pipes running to "the beach." George turns a small caffe latte-type spigot to show me the raw crude, a light-colored mix of oil, water, and gas that he lets run over his fingers. BP, which used to dump its processed water over the side, now reinjects it into the wells to keep the head pressure up.

During the 24 hours before I arrived, Pompano produced 46,641 barrels of oil, 63,887,000 cubic feet of gas, and 15,692 barrels of subseabed water.

The bottom of the Gulf is spider-webbed with 33,000 miles of pipes . . . , along with underwater well heads and production complexes.

Along with wells drilled from the platform, Pompano also has a "tieback" pipeline to eight sub-sea oil wells in 1,850 feet of water, 4½ miles away, that were drilled and installed by ship. A new platform under construction will have a 30-mile sub-sea tieback. These tendril-like tiebacks represent a new trend toward remotely controlled operations, with much of the work of surface platforms (like separating the oil, gas, and water) now taking place on the seafloor.

Having shown me the drill deck (living quarters) and production deck, which also houses the electric generators (the platform operates on 3.2 megawatts of power), George takes me down to the sub-

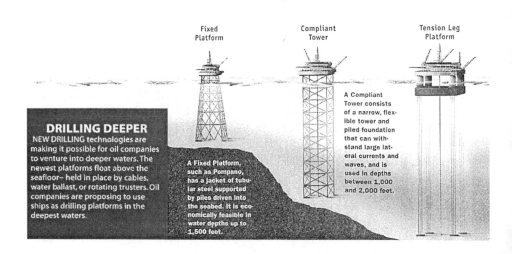

DRILLING DEEPER
NEW DRILLING technologies are making it possible for oil companies to venture into deeper waters. The newest platforms float above the seafloor– held in place by cables, water ballast, or rotating trusters. Oil companies are proposing to use ships as drilling platforms in the deepest waters.

Fixed Platform

A Fixed Platform, such as Pompano, has a jacket of tubular steel supported by piles driven into the seabed. It is economically feasible in water depths up to 1,500 feet.

Compliant Tower

A Compliant Tower consists of a narrow, flexible tower and piled foundation that can withstand large lateral currents and waves, and is used in depths between 1,000 and 2,000 feet.

Tension Leg Platform

cellar where the fire pumps, hydraulics, and utility equipment are located. Here I get to check out the large gray pipes that drop to the seabed before carrying the oil and gas ashore. The bottom of the Gulf is spider-webbed with 33,000 miles of pipes like these, along with underwater well heads and production complexes.

A platform like Pompano costs around $350 million to build and operate, Depland tells me. On the horizon we can see an even more expensive platform: Chevron's $750 million Genesis Spar, which is supported by water ballast and mooring lines, and operates in 2,600 feet of water.

That evening we eat a tasty dinner, including jambalaya, crawfish étouffée, corn bread, french fries, and ice cream. Rig dining may not be heart-healthy but at least none of the crew, out here appear to be undernourished. Later we go to sleep in double-stacked steel shipping containers converted to crew quarters.

The next day, we take the 15-mile helo-hop from Pompano over to Amberjack. Rigs, I discover, are named after their lease sales, which, for security reasons, are given theme-based designations by the oil companies' secretive exploration departments. That way, if some drunk is overheard in a Houston bar mentioning how many millions his company bid on Bullwinkle, it won't mean anything to an eavesdropper. Lease sale themes have included rock bands, country-and-western singers, types of booze, game fish, and, as with Bullwinkle, even cartoon characters.

Amberjack is the ultimate Tinkertoy. An active drilling rig, it towers 272 feet from the waterline to the top of its bottle-shaped derrick. With a four-story metal crew building, helipad, flare-off tower, tanks, processors, compressors, drill deck with 8,300 feet of piping stacked 12 feet high, 1,000 barrels of drilling mud, mud shakers, cement, two big yellow cranes, an office shack, lifeboats, and hundreds of other pipes, tubes, racks, gears, lines, and computerized systems hanging out over both ends of its legs on thick steel shelves, Amberjack is a structural salute to human ingenuity.

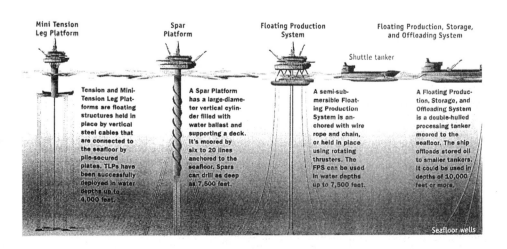

You know whoever designed this thing doesn't waste closet space at home. Still, from the air Amberjack looks small and somewhat fragile set against the whitecapped expanse of the Gulf's blue waters.

The winds are howling close to 40 knots today, the swells are around 12 feet, and with an extra half-million pounds of drilling gear onboard you can feel some sea movement on this platform. Once inside, we're given a safety lecture and told to remove rings and watchbands to avoid "degloving injuries," in which the skin and meat can get ripped off your hands. Then we meet Cary "Call me Bubba" Kerlin, the red-faced, somewhat spherical drilling supervisor.

"Might look a little dirty," he warns us. "We've been getting a lot of gumbo mud while we've been drilling." Gumbo is a heavy, thick, gray-black mud that's hard to wash off.

As a drilling supervisor or "company man," Bubba has been around the oil patch, having worked in Colombia, California, and Alaska, as well as in the Gulf. Under Bubba is the tool pusher, or rig manager. Then there's the driller who controls the drilling console, the skilled roughnecks who work for him, and the less skilled roustabouts or general assignment workers. There's the mud man, or fluids engineer, who oversees the lubricating muds (polymers, clays, dirt, and additives) that circulate down the pipe string, and several stories above the mall stands the derrick man on his monkey board, a small catwalk from which he handles the high end of 42-foot sections of pipe. As they tilt up toward him, he leans out almost horizontally in his harness to grab the top of each pipe and align it with the heavy rubber fill-up tool that adds drilling mud to the pipe string.

Bubba takes me up to the drill deck. It's a noisy, thrilling scene—a choreographed dance of steel pipe, muscle, and machine. The cranes lift the pipe to the roughnecks and roustabouts in their hard hats and steel-toed boots, and they manhandle it into position below the derrick with its massive yellow top-drive and block. One of the crew has a T-shirt reading, "New Rig, New People, New Records." I stand near the console on the waterslick deck watching the crew working the hydraulic tongs around the pipestem and threading it into the hole with a creaky slow rotation.

Right now they're down to 8,387 feet. With more than 36 other wells already down there, they're drilling this hole at a 45-degree angle, although they're capable of drilling like a boomerang, going down and then up again.

Another 42-foot section of pipe is chain-winched onto the derrick floor like a skidder-pulled log coming up a freshly cut hillside. I move forward and begin taking pictures of the redhelmeted derrick man as he leans out from his monkey board like a trapeze artist to grab the pipe and line it up with the rubber mud hose dropping down on him from above. I'm lining up a shot when one of the roughnecks sneaks up behind me and slaps my ribs, letting out an animal howl.

I turn around quizzically. He's grinning happily. "I can't believe you did that," another guy semi-shouts to be heard. I didn't jump at the prank because I knew there were no predatory animals lurking on this rig, other than these guys, of course.

On the way back to the helicopter, I spot the crane operator on a break, standing on the catwalk outside his cab, licking an ice cream cone and staring off into the blue frontier.

While major disasters like the 1969 Santa Barbara oil spill, the Gulf of Mexico's 1979 Ixtoc rig blowout, and the 1989 *Exxon Valdez* oil tanker spill in Alaska's Prince Williams Sound are rare, every year brings less dramatic spills and discharges. Just before I visited the BP platforms, there was a "minor" spill of 47,000 gallons of oil from a Chevron pipeline near Grand Isle, Louisiana, that created a 4-mile-long oil slick and fouled a small barrier island. My first day offshore, a natural gas blowout occurred at the Apache Well platform, resulting in two injuries and a great deal of damage to the drill deck. A month later, a 94,000-gallon oil spill created a

---

> *I wonder about the effects of future oil spills that will inevitably occur in deep water, just as they have following every earlier industry innovation.*

7-mile-long slick 75 miles offshore when a drilling rig anchor was dragged across an underwater pipeline.

According to figures from the U.S. Department of the Interior's Mineral Management Service (MMS), the agency responsible for leasing drilling tracts to oil companies, the 1990s saw an average of 243,650 gallons of spilled oil fouling the Gulf's waters every year. Spills can look like anything from a rainbow sheen on the water and tar balls on the beach, to a thick sludgeline turning bayou swamps into sticky asphalt. A major accident such as the Ixtoc blowout created a fiery cauldron of bubbling gas, oil, and twisted wreckage that burned for 10 months, killed several divers working to control it, and left a wide band of black tar along hundreds of miles of Texas shoreline.

I wonder about the effects of future oil spills that will inevitably occur in deep water, just as they have following every earlier industry innovation, starting with the first offshore wells built on wooden piers near Summerland, California, in 1896. Little is known about conditions that exist in the depths of the Gulf below 8,000 feet. Last fall, a deep diving expedition using the Woods Hole Oceanographic Institution's submersible *Alvin* discovered powerful "abyssal storms" scouring the bottom at 1.5 knots (the typical currents in deep water are only about one-tenth of a knot). Along with this never-before-witnessed phenomenon, the divers discovered

RESEARCH SPENDING   WHAT will be
the next big energy source? Where government
research dollars flow provides a clue. While
President Bush's proposed budget slams solar and
wind, coal is a clear winner. Fusion funding also
stays strong, even though we're at least two
decades away from a working plant. Here's where
the money's flowing:

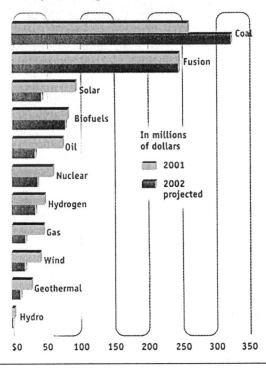

unique plants and animals living near benthic gas seeps and deep-
sea mats of marine bacteria. Recent research has also found that
the Gulf's deep waters attract hundreds of endangered sperm
whales.

No one knows much about what happens when oil and gas are
released in this extremely cold, high-pressure environment. "It's a
frontier area," admits Bob LaBelle, chief of the Environmental Divi-
sion at MMS. "Finding, tracking, and remediating [spilled oil] will
be very challenging. Where it will surface, when it will come up, and
what it will look like are all big questions."

To answer these questions, MMS and 23 oil companies have created a Deep Spill Joint Industry Project. Established in December
1998, it has spent about $2.5 million to date—about the value of two
and a half days' production from the Pompano platform—trying to
develop computerized spill models and doing a test release of oil and
gas in deep waters off Norway. Norwegian scientist Oistein

Johansen predicts that the plume from a deep-water blowout would surface hours after the disaster, and miles away from the site, and would be so widespread as to be uncollectable.

While all the test results have not yet been made public, it's clear that no environmental data are likely to hold back the deep-water energy boom now taking place in the Gulf. Since former Secretary of the Interior James Watt created MMS in 1982, the agency has never canceled a lease sale based on an oil spill risk assessment. "It's hard to make or break something as big as a lease on one issue," explains LaBelle.

Although Gulf rigs are headed for deeper water, plans for expanding deep-water drilling in Florida have sparked widespread opposition, including from the president's brother, Florida Governor Jeb Bush. Climatologists and ecologists argue that it's time to begin a shift to cleaner energy technologies that, unlike carbon-rich fossil fuels, don't contribute to global warming.

BP's CEO John Browne was the first leader of a major oil company to acknowledge that the science on climate change is sound. Looking to the future, BP has begun calling itself an energy company rather than an oil company, and BP officials have run ads saying the company's name (an abbreviation of British Petroleum) now stands for "Beyond Petroleum."

"We see a shift to lower carbon energy," BP's Hugh Depland explains. "The world is moving from coal to oil to natural gas and then to carbon-free hydrogen." Looking out at the rigs scattered across the Gulf, he acknowledges this won't be an easy transition.

As technologically challenging as the search for deep-sea oil and gas has been, the shift to new energy sources such as photovoltaics, wind turbines, biomass, and fuel cells will make today's oil-patch innovations pale by comparison.

For now, however, the Gulf's roughnecks and roustabouts continue to practice their dangerous and challenging craft with the same professional pride as America's 19th century whalers, who by extracting leviathans' living oil, lit and lubricated an earlier industrial age, until they too passed into history.

Coal Gets Cleaner—and Better Connected[3]

By Patrick A. McGuire
Business Week, May 28, 2001

East of Tampa in rural Polk County, Fla., there's a $300 million state-of-the-art power plant that serves as Exhibit A in the case for "clean-coal" technology. The plant combines two technologies never brought together before. In the first, coal is converted into a cleaner, synthetic gas. Next, the heat is recycled. Put together, the technologies make Polk's generators up to 12% more efficient than last-generation plants. And that means more power from less coal. All told, this 250-megawatt showcase plant emits significantly less sulfur dioxide, nitrogen oxides, and other pollutants than conventional coal plants.

Engineers have spent billions of dollars of government and industry money over the past decade to develop such clean-coal technologies. These methods, they say, will allow coal—which now provides slightly more than half of the nation's electricity—to continue to play an important role in U.S. power generation without unduly polluting the environment.

Environmentalists and other critics shudder at the very mention of "clean" coal, arguing that it can never be made acceptable from an environmental standpoint. Extracting coal from the ground is a messy, polluting process. And burning it, in no matter what form, always releases carbon dioxide, a greenhouse gas. "The term 'clean coal' is a myth, a very cynical term," says Lexi Schultz, a lawyer at the U.S. Public Interest Research Group. The advocates' goal, she says, "is to increase the use of coal, period."

That criticism isn't likely to go far in Washington, however. Clean coal's advocates have a powerful friend in President George W. Bush, who was the No. 1 recipient of coal-industry campaign contributions, according to the nonpartisan Center for Responsive Politics. Environmentalists say that his ties to the coal industry partly explain his February decision to withdraw U.S. support from the Kyoto Protocol, the international agreement to curb global warming by restricting carbon dioxide emissions from the burning of coal and other fossil fuels.

The Administration is also considering requests from the coal and utility industries to drop Clinton Administration–initiated lawsuits that would require new pollution controls to be installed on old coal-

---

fired power plants. And Bush has shown no enthusiasm for a Senate bill that seeks to tighten restrictions on coal plant emissions and require older power plants to meet stringent Clean Air Act standards. Instead, the President has incorporated into his budget proposal $2 billion in federal funding to promote clean-coal technology.

More Efficient. Tampa Electric's Polk Power Station relies on some of the latest advances in a field of research that began in the 1980s, when coal plants were shown to contribute to acid rain. In 1990, Congress amended the Clean Air Act,

> *Improvements in antipollution "scrubbers" . . . have reduced emissions of sulfur dioxide by 95% or more.*

phasing in tighter limits on emissions of soot particles, nitrogen oxides, and sulfur dioxide from power plants. To help implement those limits, the Energy Dept. has since spent $1.8 billion on 39 clean-coal technology programs, while state governments and industry kicked in a further $3.4 billion.

Several of the resulting technologies are already in wide use. They include new equipment that cleans coal after it is mined, new methods of burning coal, and advances in treating gases after burning. This so-called reburn technology splits the burn into two stages for more complete combustion, which results in lower emissions of nitrogen oxides. The technology has already been adopted by three-quarters of the nation's coal-fired plants. Meanwhile, improvements in antipollution "scrubbers," designed to treat exhaust from smokestacks, have reduced emissions of sulfur dioxide by 95% or more.

In recent years, much of the research has focused on the development of more efficient burning methods so that more electricity can be extracted from less coal. An average pound of coal contains about 10,000 BTUs. Existing coal-fired plants capture only 33% to 35% of that. The rest is wasted. But plants built in the past five years, such as the one in Polk County, can extend that slightly, extracting 37% of the energy from a pound of coal. And those being built now, using the Polk plant combustion technology or a similar method, are expected to have efficiencies of about 40%.

Little to Fear. In the future, clean-coal enthusiasts anticipate even more dramatic strides in efficiency and environmental quality. An Energy Dept. projection for coal sets an efficiency goal of 60% by 2025. On the environmental front, the trick, says Thomas A. Sarkus, senior engineer at Energy's National Energy Technology Laboratory, is balancing environmental measures with affordability. "It's not a technical matter," he says. "The more controls you want, the more it's going to cost."

The coal industry fears those higher costs. They could eliminate coal's biggest selling point—that it's cheaper than natural gas— thus forcing utilities to switch to other fuels, including gas. But so far, such a shift seems unlikely. "What's the alternative to coal?" asks Ronald H. Carty, director of the Illinois Clean Coal Institute. "If you're talking renewables, I don't know how you do it."

In fact, coal executives have little to fear from alternative energy. While coal provides more than 50% of the nation's electricity, alternative sources, such as solar and wind power, provide less than 3%. The percentage of BTUs generated from coal is expected to drop slightly in the next 20 years, says Sarkus, but demand for power will grow so much that coal usage, in tonnage, will jump 30%. Technologies such as those used in the Polk County plant in Florida are the only way to avoid the potentially harmful environmental consequences of that surge in coal use. Clean coal is clearly here to stay.

There's No Recession Among Roughnecks on Natural-Gas Rigs[4]

BY ANN ZIMMERMAN
WALL STREET JOURNAL, SEPTEMBER 7, 2001

When a football scholarship fell through, Bronson "Bull" Holmes dropped out of college. The tuition was just too much of a burden on his single mother, who runs a convenience store.

Standing on the grimy, hot deck of Rig 203 as it corkscrews through 20,000 feet of mud and rock, the 26-year-old derrick hand says he always meant to go back to school to study criminal justice. But with drilling busier than it has been in years, things are just too sweet for him to leave this town 1 1/2 hours north of Houston.

So sweet, in fact, that three months ago he paid off the mortgage on his mother's mobile home. In the past two years, he has had three big raises. He now pulls down nearly $50,000 a year as the drilling company he works for fights a serious labor shortage in the oil patch.

"We ain't hurtin' now," says Mr. Holmes.

San Francisco and San Jose may have been the places to be two years ago. But in the midst of the broadly deteriorating U.S. economy, employment boom towns have been springing up in places such as Rock Springs and Gillette, Wyo., across east Texas and the Rocky Mountains. Rigs have sprouted all over as more natural-gas wells are being drilled than at any time in nearly 20 years. This summer, the number of active rigs hit 1,293, more than double the number operating in early 1999—darker days for the industry.

Applicants here don't need fancy degrees or venture capital to be roughnecks, the term for lower-level oil-field workers. Rather, muscle, the ability to survive sweltering heat and a willingness to bunk with co-workers can pay off. Drilling companies and contractors are in such desperate need of hands that they are paying $75,000 a year or more to roughnecks working seven days on and seven days off.

Two years ago, there were hardly any such job openings and they paid as little as $20,000. But natural-gas prices went from a low of $2.25 per thousand cubic feet in late 1999 to $3 in the spring of 2000, soaring to a high of $10 in late 2000. Though the price has fallen back recently to less than $3, new gas-fired power plants are due to come online in the next few years, and demand for gas is likely to rise.

High prices have meant fabulous profits for energy companies and the businesses that serve them. Nabors Industries Inc., the Houston driller that employs Mr. Holmes, saw its profit more than quadruple, to $187 million in the first six months of this year over the like period in 2000.

Some of those newfound oil riches are helping to rejuvenate once-sleepy towns such as Fairfield and surrounding Freestone County. Between the energy boom and a new power plant going in, the streets of Fairfield, a town of 3,500, are clogged with traffic. Restaurants are packed; rental housing is impossible to find. The Holiday Inn Express, one of four motels, has been booked solid on weekdays for about a year, according to hotel manager Jay Patel.

> *Between the energy boom and a new power plant going in, the streets of Fairfield, a town of 3,500, are clogged with traffic.*

Business has been so brisk at the Bossier Country Chevrolet and Chrysler dealership that it recently completed a $2 million remodeling and expansion, aided by a 15% increase in gross sales. And thanks to natural-gas drilling, the value of taxable property swelled by $500 million, or 42%. The tax rate has been reduced and, even so, tax revenue is up $5.3 million, or 30%, this year.

County employees were given a 10% raise, county Judge Linda Grant says, and plans are afoot to fix streets, double the appropriation for renovating the Freestone County Museum and give $15,000 to a railroad museum.

The renewed activity in the oil patch has been a boon to Gilbert Daniel, owner of five restaurants in town. He estimates sales are up 20% this year.

Yet to the people around Fairfield, this boom seems different from the last one, tamer and somewhat less ostentatious. Sheriff Ralph Billings chalks it up to folks learning their lesson in the early '80s. "After going through two busts, it seems more people aren't living so wild and free this time," he says. "Now there is a realization, from top to bottom, that good times can end just as quickly as they came."

The workers doing the heavy lifting say they are trying to be careful with their money this time—not running off to Louisiana casinos so much.

The roughnecks put in 12-hour days, at wage rates ranging from $13 an hour to $19.25. For more than half their time, they are paid time-and-a-half. They also collect a $20-per-day food allowance. Most of the men live too far away to commute, so after their shifts, they kick back in rent-free, air-conditioned mobile homes, with full kitchens and a washer and dryer, sharing a room with six other guys in bunk beds.

The industry lost one generation of hands in the mid-1980s slump and another in the slump in late 1998, says James Nash, a burly, snuff-dipping Nabors superintendent sporting mirrored sunglasses and a gold-nugget ring.

When work started picking up last year and the number of rigs in and around Fairfield grew, Nabors put up Mr. Nash, his wife, Brenda, and their two toy white poodles in a doublewide trailer a few miles off the highway. They check on their house in Nacogdoches, about 100 miles away, every few weeks.

On the wide front porch of the mobile home, which doubles as a field office, a pile of job applications sits on a table under a paperweight that says "24-7." Most days, he gets pretty far into the stack. Mr. Nash has hired about 400 workers in the past year to work here in the Bossier Sand Pla oil and natural-gas field.

"It's a gravy train now," says Nathan Simmons, Rig 203's tool pusher, the equivalent of a ship captain. Mr. Simmons, now earning about $75,000 a year, just bought himself a brand-new Ford 350 diesel dual-exhaust truck, in silver, and a travel trailer. His wife, a nursing director for a home health agency, just got a new Buick.

The work can be hard and potentially dangerous, and it takes him away from his wife and three children, who live 3 1/2 hours away on a farm in Louisiana. But he says, "I never dreamed I'd be making this much working six months a year."

Such a dream is what prompted Royce Bayless to hitchhike five days from Joplin, Mo., in search of a job. He had just left his marriage of 15 years—"It wasn't working out," he says—arriving with the clothes he was wearing and a knapsack with some extra shirts and a razor.

Mr. Bayless, 36, was a derrick hand for Nabors in Wyoming in the mid-1990s before the work dried up and he ended up peddling fruit in Missouri. Once his previous experience was confirmed and he passed a drug test and physical, Mr. Bayless was given a job as a floor hand screwing together 34-foot-long drill pipe on Rig 255.

There, he started working for tool pusher David Roop, 37, who recently bought a 22-foot Cajun fishing boat and purchased 6 1/2 acres of land in the bucolic Texas Hill Country—the same part of the state where dot-comers and "Dell-ionaires" from Austin snapped up acreage. There, Mr. Roop and his wife plan to build their dream house.

Mr. Bayless has dreams of his own. "I'm ready for a new start, a new stake in life," he says. "And I knew I'd find it in the oilfields."

II. Energy Alternatives

Renewable Energy Consumption in North America, 1980–2020

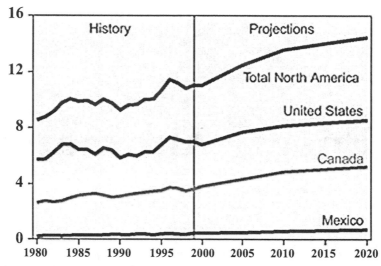

Source: History: Energy Information Administration (EIA),
Office of Energy Markets and End Use, International Statistics Database
and *International Energy Annual 1999*, DOE/EIA-0219(99)
(Washington, DC, January 2001). Projections: EIA, World Energy
Projection System (2001).

Editor's Introduction

P eople have long recognized that population growth and the expansion of human activities have taken a toll on the environment. As far back as 1306, Kind Edward I of England issued a call to ban a certain type of coal due to the foul smoke it produced; in the U.S., the Air Pollution Control Act of 1955 was the first in a series of federal laws passed to improve the quality of air—much of it befouled by the burning of coal. The environmental movement, which gained momentum after the first Earth Day in 1970 and the first United Nations conference on the environment in 1972, has led the call for cleaner energy sources. Then, in 1973, the Organization for Petroleum Exporting Countries (OPEC) dramatically raised oil prices and led an oil embargo against the U.S. and the Netherlands to protest the Western nations' support of Israel. The price of crude oil skyrocketed, from $3 a barrel in 1970 to $30 a barrel in 1980. After drastic oil shortages and a global economic recession, it became clear that a reliable, sustainable supply of energy was crucial to the world's stability—as well as to its environment.

The articles collected in this section assess the viability of alternatives to traditional fossil fuels, including wind, solar, geothermal, biomass, hydroelectricity, and other "renewable" sources. While once considered prohibitively expensive, the costs of wind and solar power generation, in particular, have plummeted in recent years (After factoring in the massive subsidies given to fossil fuel industries, renewables seem even more competitive.) Renewable fuels comprise only 2 to 3 percent of America's energy supply, although the percentage is higher in many European countries. Denmark, for example, reaps about 15 percent of its electricity from wind. Nevertheless, many doubt that renewables will ever be able to produce the kind of abundant energy needed to fulfill the world's rising energy demands. Advocates of alternative energy put their faith in technology with the knowledge that, historically, the search for better energy sources has gone hand in hand with human progress.

Traditional energy companies have already begun to explore energy alternatives. Writing for a major trade publication, *Oil and Gas Journal*, Will Rowley and John Westwood analyze the market potential of renewable energy over the next 10 years. They predict that many factors will contribute to rising demands for renewables, including the passage of the Kyoto Protocol, which demands that participating countries lower their carbon dioxide emissions from the burning of fossil fuels, and the growth of developing countries, which would prefer to use indigenous, renewable sources, rather than import expensive fossil fuels.

Wind is the fastest-growing renewable energy source; according to the American Wind Energy Association, wind-propelled turbines will supply 6 percent of the nation's electricity by 2020 (today, it supplies about 1 percent). As Janet Raloff reports in "Power Harvests," American farmers have begun to erect turbines on their wind-swept fields. Such turbines—which are much more sophisticated than conventional windmills—are inexpensive to build and maintain. In the near future, wind power generation will occur mostly on small farms and ranches, though creating the infrastructure to transport the wind to energy-hungry cities will be an enormous challenge.

In "Sunny Solution," Peter Fimrite writes that solar energy may no longer be "an eco-hippie pipe dream." Improved photovoltaic products—the panels used to capture the energy of the sun's warming rays—have helped to drive down the costs of solar energy; in a sunny state like California, which supports alternative energy through legislation and tax incentives, solar is becoming a viable option.

Biomass energy is created from burning organic materials like sugar cane or corn; it's an old method, and still the main option in many poor countries, where people gather wood or dung for fuel. But, as Laurent Belsie reports, sophisticated bio-fuels are also emerging as new sources of renewable energy. Ethanol, a fuel made from corn, is experiencing a boom in demand, and bio-diesel, a mixture of diesel and soybean oil, seems to work as well as diesel and burn much more cleanly. Like the wind turbines in farmers' fields, these renewable energy products bring new revenue into beleaguered farming communities.

Amy Worden reports on "green power," a new trend that has emerged from energy deregulation. Pennsylvania is one of a handful of U.S. states where energy monopolies have been dismantled and energy suppliers have been allowed to compete on the open market. Given the choice, nearly 20 percent of the state's consumers have been willing to pay more for energy that is produced, in part, from renewable resources. The Green Mountain Energy Company, which offers consumers a range of renewable energy mixes, is betting that this market will grow; indeed, by 2002, 20 states are scheduled to enact some form of deregulation.

Two articles in this chapter look at "micropower"—power created through independent, off-the-grid systems. New technologies such as wind turbines and fuel cells would allow individual homes and businesses to produce their own power—instead of tapping into a power grid whose energy source may be hundreds of miles away. As Lucy Chubb reports for the *Environmental News Network*, micropower would benefit high-tech businesses, which need a guaranteed electricity flow, as well as developing countries, where people in remote areas often cannot access an energy grid. The accompanying *Wired* article elaborates on the inefficiency of today's power grid, in which a majority of energy created through burning fossil fuels is lost forever—as waste heat that goes into the atmosphere, or in progress to your door.

In this section's last article, Robert Ayres argues that conservation is the most practical and effective means of reducing greenhouse gas emissions. The best way for an industrialized nation like the U.S. to conserve is to improve the efficiency of its energy generation: according to Ayres, the U.S. currently loses 19 out of every 20 energy units it produces.

Hype Aside, Renewable Energy Emerging as Viable Long-Term Energy Industry Business Opportunity[1]

By WILL ROWLEY AND JOHN WESTWOOD
OIL & GAS JOURNAL, OCTOBER 22, 2001

Renewable energy development is beginning to emerge as a viable, long-term business opportunity for energy companies.

The subject of renewable energy presents the average businessperson or investor with a number of difficulties.

First, it is often associated with "new age" quirky thinking, when in reality renewables are the oldest and most proven form of energy generation.

Second, the subject suffers from overly extravagant resource-based claims that are far from realizable. (For example, the global wind resource is vast; however, the proportion that can be economically captured for power generation remains small.)

Third, it is full of inconsistency—in definition, description, and reporting.

A key aim of this article is to clarify the situation and provide a summary of the significant business opportunities likely to be presented over the next decade. In order to do this, we have examined each resource, the associated technologies for its exploitation, present levels of capital expenditure, and then developed capital expenditure forecasts by sector. After stripping away the hype, what remains is a major industry that is developing strongly and offers considerable long-term business opportunities.

Drivers

Four key factors are driving the rapidly accelerating development of the renewable energy industries:

- Environmentally driven markets. These are generally the Organization for Economic Cooperation and Development countries that, in order to meet mandated emission targets emanating from the United Nations' Kyoto Protocol on climate change, will sharply accelerate their use of renewable energy. These markets are expected to provide the largest gain in the

---

use of renewables in the short-term and the largest short-term business opportunities.

- Energy-driven markets. These are particularly the Asian and other developing economies where demands for new energy are being propelled by population growth, industrialization, and urbanization. Government and industry are predisposed to meet new electricity requirements from renewable sources, which are indigenous and have relatively short installation timeframes. The energy demand from these markets is expected to overtake that of the OECD countries beyond 2002 and become a significant sector by 2007.

- The green consumer. The liberalization of energy policy and the rise of the political and "green" consumer are also driving greater demand. The breakup of state energy monopolies has spawned competition and allowed businesses and individuals to choose their provider of electrical services. In many markets, the options for industrial, business, and residential consumers include "green energy" programs. While still a minor part of the mix, green energy options will gradually increase awareness and attention and cause increased market demand from some customer segments.

- Economic drivers. The market is increasingly being influenced by traditional market drivers rather than by political forces. In some situations wind, biomass, and small-scale hydropower are becoming viewed as complementary to fossil fuels, not as a replacement, and there is also an observable trend among conventional energy producers to consider the use of renewable technologies.

Wind Energy

Over the past 6 years, wind energy has been the fastest-growing renewable energy source,[1] with additional output growing at a compound rate of 30%. It is likely to maintain this position for the medium term, as some projects have now achieved output and economy ratings comparable with fossil and nuclear power.

Europe and North America have dominated wind energy developments over the past 2 decades. However, the past 5 years have seen the European region forge ahead in terms of installed capacity, and with the current economic and political situation, it is unlikely that this situation will change. While currently accounting for 73% of installed capacity, European growth has slowed slightly to 36%/year from over 40%/year. However, while other regions have grown faster, the large installed base of Europe has maintained its preeminent global position.

One interesting situation has been the development within Western Europe of offshore wind farms. Western Europe has a higher average population density coupled with a higher offshore wind resource than North America. This has made offshore wind a viable alternative for Denmark, Ireland, Sweden, Norway, Germany, and the U.K. There is currently about 80 Mw of installed capacity offshore, all of it in Europe, with Denmark responsible for 63% and the Netherlands 16%.

There are, however, extensive development plans throughout the region that could lead to a massive increase in installed capacity by 2010. Among the U.K., Denmark, the Netherlands, and Germany, there is the

> *Over the past 6 years, wind energy has been the fastest-growing renewable energy source.*

possibility of nearly 6,500 Mw of installed capacity, which, if realized, would make offshore wind the fastest growing sector of the wind energy market.

In the U.S., huge tracts of sparsely populated land with good wind potential and suitable grid connections have made onshore wind generation the economic option.

Over the next few years, continued technological improvements, such as the development of 5 Mw turbines, should lead to further efficiency and output gains and see wind energy firmly ensconced as a commercial proposition for large-scale, subsidy-free power generation.

Manufacturing wind turbines is becoming a significant business in some countries. For Denmark, an early pioneer in the sector, this has become the country's largest engineering export.

Solar Energy

Solar energy is becoming a well-established business sector. The technologies can be characterized into two distinct groups: photovoltaic (Pv), or "active," and thermal, or "passive." Pv directly converts light to electricity, while passive systems use solar energy for direct heating or to generate electricity indirectly through a heat-transfer system. The passive use of solar energy is separated into solar heating and solar thermal electricity.

With 1 sq. m. of installed solar collectors saving 200–600 kw-hr/year of electricity—depending on the country, the type and quality of the installation, the weather conditions, and consumption pattern—it is projected that the solar collectors to be installed in Europe (mainly southern Europe) by 2005 will produce 14 million Mw-hr of solar thermal energy.[2]

There is no global estimate of the number of passive systems in use today, but there are an estimated 5 million plus systems in use in developed nations.[3]

The major energy companies BP, PLC, and Royal Dutch/Shell Group are both significant players in the Pv solar technology sector.

Tidal, Current-stream Energy

Engineers have been trying for hundreds of years to effectively harness what is widely acknowledged as one of the great untapped energy resources of the planet—the energy associated with the movement of ocean tides and current streams.

However, most potential sites are poorly located with regard to prospective users and often in areas of high ecological value. But technological developments of the last decade coupled with the growing acceptance of the need for renewable energy should ensure that tidal and current-stream energy finally move out of their development stage and into ccommercial reality.

One extreme example is a 2.2 Gw tidal fence being planned by Blue Energy Canada Inc., Vancouver, B.C., for the San Bernadino Strait off the Philippines. The project, estimated to cost $2.8 billion and take 6 years to complete, would be the largest tidal power plant in the world.

However, the main future of tidal and current stream energy schemes may lie in small-scale tidal power plants similar to the new generation of 3–5 Mw wind turbines that are being adapted to subsea use. With lower capital costs and a close matching of output to local demand (removing the cost of transmission), small-scale tidal power could have a very bright future, particularly in South America and the Asia-Pacific region, where grid connection is a major structural issue.

Wave Energy

In the wave energy sector, a number of devices have been designed and deployed, but no one device has yet to reach either universal acceptance or widespread commercial use.

The last decade has, however, seen significant developments in efficiency and reliability that have brought wave energy from a conceptual to a development phase in its progress to commercialization.

There is also growing local, national, and international support for the fledging industry that should lead to competitive commercial applications before 2010.

The U.K. and Japan have dominated the development of wave energy devices over the past 15 years and currently account for over 60% of worldwide development expenditures.[4]

The U.K. industry has recently been boosted by the announcement that Europe's largest dedicated research facility is to be built at Blyth in Northeast England.[5] Containing a large converted dry dock, the facility is capable of testing full-size devices of up to 30 kw.

Geothermal Energy

Perhaps the best-known example of geothermal energy use is in Iceland, where it is the second-largest source of energy. Reykjavik, with a population of more than 145,000, pipes hot water to every house at a cost lower than cold water.

The global potential is spread quite evenly across the regions, although realizable prospects are likely to be concentrated in Europe, Asia, and North America. However, as a result of an extensive research and development program, the U.S. has the largest quantifiable resource.

Hydrothermal reservoirs are large pools of steam or hot water trapped in porous rock. To create electricity, the steam or hot water is pumped to the surface, where it drives a turbine that spins an electric generator. Because steam resources are rare, hot water is used in most geothermal power plants.

Steam and hot water power plants use different power production technologies. A standard geothermal power generation plant may have 6–10 producing wells and 1–4 injection wells costing $1.5–3 million each.

In the U.S., there are subsurface volcanic hot spots under Yellowstone National Park and Hawaii and intraplate extension with hot springs in the Great Basin of the U.S. West.

California generates the most geothermal electricity, with about 824 Mw at the Geysers complex in northern California (much less than its capacity but still the world's largest developed geothermal field and one of the most successful renewable energy projects in history), 490 Mw in the Imperial Valley in southern California, 260 Mw at the Coso resource in central California, and 59 Mw at smaller plants elsewhere in the state. There are also geothermal power plants in Nevada, Utah, and Hawaii.

Due to environmental advantages and low capital and operating costs, direct use of geothermal energy has skyrocketed, with over 450,000 geothermal heat pumps installed. In the western U.S., hundreds of buildings are heated individually and through district heating projects with geothermal energy.

Biomass

It is estimated that biomass accounts for at least 15% of total world energy, and in some developing countries this figure rises to 35–50% of domestic energy supply.

The conversion of biomass for electric power generation is often referred to as "bio-energy." Modern biomass usually involves large-scale plants and aims to substitute for conventional fossil fuel energy sources. It includes forest wood and agricultural residues, urban wastes, and biogas and energy crops.

Traditional biomass is generally confined to developing countries and small-scale uses. It includes fuel wood and charcoal for domestic use, rice husks, other plant residues, and animal dung. Industrial applications include combined heat and power, electric power generation, space-heating boilers in public buildings, domestic heating boilers, and decentralized energy applications.

Biomass plants have been installed across the world, burning a wide range of renewable fuels. India has a national program of typically small, 5 Mw plants. Plants in the U.S. are burning wood processing residue and in the U.K., straw and poultry litter. The U.K.'s first—and the world's largest—straw-fired power station at Ely was built for capital outlays of $90 million and has an output of 36 Mw.

While the North American market will remain significant, South America is likely to contribute most of the biomass energy growth in the Americas. Indonesia, India, and China will drive forward the Asian market for biomass electric power generation.

Small-scale Hydro

Large-scale hydro is a mature industry that, while involving a renewable energy resource, is also generally regarded as having unacceptable environmental consequences. However, the small-scale hydro market is a different matter. Small-scale (<10 Mw) hydro schemes normally involve a diversion of some of the flowing water rather than a complete holdback-and-discharge format of larger schemes. From being one of the earliest forms of power generation, it has for the past 30 years been overshadowed by the development of large-scale hydro schemes.

Improvements in turbine and generator efficiencies, particularly over the past 10 years, have seen small-scale hydro grow into a distinct industry. The extent of this transformation can be highlighted by the active involvement of major turbine manufacturers such as General Electric Co.

The small-scale hydro market has grown over the past decade to become a $2 billion/year industry. Through 1995–2000, small-scale hydro has been reassessed in many of its major markets, and new financial incentives have been put into place that are leading to increased interest and activity.

Asia—in particular China—is the backbone of the small-scale hydro industry, representing nearly 50% of expenditure, a situation that is likely to remain through 2010.

Assessing Prospects

One of the major problems in assessing future investment in renewables is that, rather like the oil and gas industry, the number of "possible" projects is considerable, but in reality, many will never be realized, are totally uneconomic, or are the subject of indefinite delay.

A prime example of this is wind power, where the central overriding issue, as with many energy sources, is not performance but the level of practical exploitation and hence commercialization. Many factors can influence this: location, available technology, topography, grid connection, visual amenity, etc., as well as less quantifiable issues, such as public perception. Grid connection is a particular problem, as some of the world's best prospective sites are a considerable distance from existing transmission grids.

A number of estimates have been made of the worldwide exploitable resource of wind energy, all of which have become underestimates within a very short period of time as technological development has made commercialization more feasible at lower wind speeds and in more locations.

The technically exploitable wind energy resource was estimated in 1998 at 53,000 Tw-hr/year,[6] slightly less than four times the 1998 world electricity consumption of 14,396 Tw-hr/year. This calculation was a composite of regional and national estimates from a number of agencies. It has been interesting to note that, almost without exception, every detailed local or regional analysis completed since then has estimated its resource at anything up to five times this World Energy Council estimate. This had led to speculation by a number of academic sources that the exploitable resource could exceed 200,000 Tw-hr/year,[7] more than 13 times 1998 global electricity consumption and nearly 10 times the International Energy Agency projection of electricity consumption in 2020, 24,000 Tw-hr/year.

Capex Outlook

Douglas-Westwood's models indicate that, during 2001–10, the total world renewable energy market capital expenditure will grow from $15 billion/year to $40 billion/year, an average annualized growth rate of nearly 10%. We expect the capex over this period to total $272 billion.

Through 2004, we expect continual growth in output across all the renewable energy sectors. However, we forecast that the capex from 2005 will, for a brief period, effectively level. We believe that a reappraisal of the cost of "green" electricity and a reduction in the renewable energy markets ability to command premium prices will temporarily slow the rate of growth.

The leveling off in the market is likely to be temporary, and as we approach the end of the decade, the demand for electricity in the Asian region, particularly China, will act as a further spur to the renewables sector. The period of 2001–05 can be characterized as environmentally driven, while 2005–10 will be energy demand-driven.

Although there are variances in the forecast growth across the different sectors, the overall trend in expenditure is upward. With a higher ratio of planned and possible projects, wind and biomass are forecast to represent the most significant proportion of expenditures. With a combined expenditure during 2001–10 of $190.5 billion, these two sectors are forecast to represent over 70% of renewable energy industry spending.

Conclusions

Renewable energy is a significant global industry with an annual capital expenditure of over $15 billion. Contrary to the belief of many, the majority of this expenditure is for conventional industrial and power generation equipment, such as turbines, generators, switchgear, control systems, cabling, monitoring systems, etc.

Major companies such as ABB Group, GE, and Paris-based Alstom have been increasingly active across many of the renewable sectors and quietly building market share. The supply of commercial off-the-shelf equipment by companies such as these has been one of the key fundamentals in reducing the cost of renewable energy.

Many equipment suppliers now view the industry not as an embryonic niche but as a "normal" business sector. This step change in approach also means that much of the technological development within the industry is now coming from industrial research and development rather than academia. It is the engineers that are now putting a commercial reality into the world of renewable energy.

The commercialization of renewable energy is gradually feeding into the energy industry's planning process. It has been noticeable that energy companies are not only more optimistic about the renewables sector but that medium- and long-term cross-sector planning is developing.

The industry is moving away from a project-by-project basis and into output planning within an integrated strategy.

Financiers and investors who are looking for a balance of projects and cross-sector risk management are also seeking to include renewables within their investment portfolios.

Notes

1. World Energy Council.
2. European Union.
3. Douglas-Westwood computation of national statistics.
4. Douglas-Westwood International Ocean Systems, July–Aug 2001, pp. 26–27.
5. World Energy Council.
6. World Energy Council.
7. Energieonderzoek Centrum Nederland.

Power Harvests[2]

By Janet Raloff
Science News, July 21, 2001

During the Vietnam War, Daniel Juhl toiled as a missile-guidance technician. But when he left the service, he says, "there wasn't a lot of call for my training. So, I thought I'd turn my spears into plowshares."

Or, to be more literal, into energy-generating wind turbines. In 1978, Juhl entered the wind-farming industry, helping design and erect small commercial systems for others—first in his home state of Minnesota, then in California, Europe, and China.

A few years ago, he decided to build and operate his own wind farm. On patches of land amounting to 6 acres, which he leased from a nearby farming family in Woodstock, Minn., he built access roads and erected 17 turbines. For the past 2 years, this operation has been generating up to 10 megawatts (MW) of electricity, depending on the winds. The local utility buys Juhl's commodity at about half of what residential customers will pay to use this electricity.

Heartened by the profitable cash flow of Juhl's operation and his turbines' low maintenance demands, the farmers from whom he leased land decided that they, too, were ready for a direct role in wind power. With Juhl's help, they're planning to install two units of their own.

Juhl expects wind turbines to be sprouting up far and wide in coming years. "Talk to some farmer for half an hour, and he'll understand what this is—just another cash crop," he says. It's not much different from reaping wheat, he notes, "except that your combines are 200 feet in the air."

Although pioneers of wind power in the mid-1970s tended to erect their wind farms on remote mountain peaks and passes, the present crop of wind advocates has begun turning to agricultural lands.

There's plenty of untapped wind to be had there. Wind mappers rank regions on their ability to produce commercially significant power using a 6-point scale. The higher the number, the greater and more reliable the wind resource. Today, "we're developing commercial wind farms in areas rated class 4 to class 6," says Greg Jaunich, president of Northern Alternative Energy in Minneapolis.

"Within 100 miles of every major metropolitan area, there's at least a class 4 wind resource," notes Jaunich. Most of these are farmed areas, which is why he and other commercial developers of environmentally attractive, or green, power have been leasing land there.

Farmers welcome the second income these offer. In Iowa, each quarter acre that a farmer makes available to a developer's turbine often with blades spanning 150 feet-can yield royalties of about $2,000 a year, notes agricultural economist Lester R. Brown, president of the Earth Policy Institute in Washington, D.C. Adds Brown: "In a good year, that same plot might produce $100 worth of corn."

Farmers who develop those wind resources themselves can reap far bigger bounties—up to perhaps $20,000 per turbine annually, Juhl claims.

Globally, wind generation of electricity has nearly quadrupled over the past 5 years, and in the United States, it's expected to grow 60 percent this year alone. As farmers struggle to make ends meet, "some are now finding salvation in this new 'crop,'" Brown observes.

As farmers struggle to make ends meet, "some are now finding salvation in this new 'crop.'"—Lester R. Brown, agricultural economist

---

"It's like striking oil, except that the wind is never depleted."

The profitability of wind power has blossomed over the past few decades. A kilowatt-hour (kWh), the basic unit of delivered electricity, is equal to the energy consumed by a 100-watt light bulb burning for 10 hours. In the mid-1970s, commercial wind turbines cranked out electricity at 30 cents or more per kWh. That was a staggering amount, considering that coalfired plants were generating it for about 2 cents/kWh.

Today, some wind turbines can generate power for less than 6 cents/kWh while utilities are, in some cases, charging customers more than twice that. Large wind farms sited where the air flow is reliable and strong can now produce electricity for as little as 3 cents/kWh—40 percent less than was possible with the best turbines a mere 5 years ago. For comparison, earlier this year, power-strapped California utilities were forced at times to buy electricity on the spot market for up to 33 cents/kWh.

Compared with more traditional—and more polluting—forms of electrical generation, wind power can be competitive economically, notes energy economist Florentin Krause of the International Project for Sustainable Energy Paths in El Cerrito, Calif. "It's dirt cheap," he says.

Indeed, he's found that the cost of wind-generated electricity is now about half the cost of nuclear power if all expenses—from facilities' construction and maintenance to demolition and disposal—are taken into account.

Solar photovoltaic electricity and other types of renewable power, he observes, typically need a substantial subsidy, such as a tax break, to even come close to competing with power from fossil-fired and nuclear plants.

What's more, Krause points out, unlike large, traditional generating stations that can take years to construct, wind turbines can be erected in 3 months—and they operate without spewing the greenhouse gases that fuel global warming.

The most impressive aspect of wind power to Randall Swisher, executive director of the American Wind Energy Association in Washington, D.C., is the magnitude of the supply. The U.S. wind-power potential, he says, "is comparable to or larger than Saudi Arabia's energy resources." In fact, Brown's research indicates that all current U.S. electricity needs could be met from wind resources

---

Wind turbines can be erected in 3 months— and they operate without spewing the greenhouse gases that fuel global warming.

---

in just three especially breezy states: North Dakota, Kansas, and Texas.

Even so, utilities have been slow to embrace the wind, and most farmers remain unaware of the value of the breezes rushing over their fields, notes Lisa Daniels. That's why she founded Windustry. The 6-year-old Minneapolis organization has provided state farmers and rural landowners, including Native American communities, with a nuts-and-bolts overview of wind's prospects and what it takes to harness that potential.

Windustry and the American Corn Growers Association, based in Washington, D.C., recently banded together to help landowners nationwide find ways to overcome the obstacles to owning the infrastructure to generate wind power.

Consider financing. From Juhl's experience, one of the biggest obstacles to small-scale wind farming is a need to "educate bankers." Unlike most other businesses, he says, wind systems "have a positive cash flow right out of the box. Each year, they produce enough 'crop' to pay the debt, to pay expenses, and to put money in your pocket." With no experience in such investments, the banks were dubious—and reluctant to issue a loan.

Independent owners of renewable-energy systems, which include wind farms, face yet another formidable challenge—negotiating with the local utility to sell their product at a profitable rate. Most small wind farmers lack the leverage and experience to cut good deals, according to speakers at last month's American Wind Energy Association's meeting in Washington, D.C.

Another disadvantage for wind farmers comes from regulatory hurdles. Big central-station power plants typically need to clear these hurdles just once to put 500 MW on line. In contrast, to get the same wattage on line, small-scale wind generators may collectively go through these transactions 200 or more times—and in as many regulatory jurisdictions.

Perhaps the biggest constraint to wind power's growth is getting the crop to market. The greatest technological need for rural wind-power development, Swisher argues, is not better turbines or electronics but "transmission infrastructure." Overcoming this, he says, "is our number one long-term priority."

Linda Taylor, Minnesota's deputy commissioner of energy, agrees. Moving electricity from rural turbines to energy-gobbling cities, she says, "is the only real sticking point for massive wind development."

It's already constraining development of Buffalo Ridge in southwest Minnesota, where winds blow steadily 320 days a year. Hundreds of turbines there are slated to deliver 450 MW of wind by the end of next year. However, Taylor told *Science News*, "we could easily get 3,000 or 4,000 MW of wind energy out of that area if we could get the transmission problem resolved." That's enough energy to power some 1 million homes.

R. Nolan Clark, an agricultural engineer and director of the Agriculture Department's Conservation and Production Research Laboratory in Bushland, Texas, sees much the same problem in his part of the country. Since most transmission lines outside of urban areas were installed in the 15 years following World War II, they are, in his words, "old and antiquated."

Sized to carry power needs of the 1950s, they're hard-pressed to satisfy the far more electricity-hungry households throughout even rural America today. As a result, many of these lines can't transmit more power, he says. Indeed, Taylor observes that any additional power fed into such lines in her state can and often does overload them. "This shuts the whole system down," she says.

These limitations highlight a major disconnect between the way power lines are configured and the new needs of small, distributed generators. An analogy with blood circulation illustrates the problem. Big trunk lines, like arteries, branch into successively smaller lines, like capillaries, which feed local areas including individual residences. Operators of distributed-power systems usually have access only to the smaller lines, although their needs require a large artery.

Upgrading rural lines would solve the problems, but at a cost of up to $1 million per mile, Swisher notes. An alternative plan might use wind to generate hydrogen on the farm, and off the grid, and then to pipe hydrogen to cities for use in automotive fuel cells.

Will Rural Winds Power Urban Cars?

"Hydrogen is the fuel of choice for the new, highly efficient fuel cell engine that every major automaker is now working on," says Lester Brown, president of the Earth Policy Institute in Washington, D.C. With Daimler Chrysler planning to roll out its first emissions-free, fuel cell-powered cars in 2003, he says, "Ford, Toyota, and Honda will probably not be far behind."

What if electricity from wind-powered turbines in North Dakota broke down water into hydrogen, which could be piped 1,600 miles to Chicago vehicles? Bill Leighty, director of the Leighty Foundation in Juneau, Alaska, presented results from a new study that projected the economics of this 2010 scenario.

Last month at the American Wind Energy Association's annual meeting in Washington, D.C., he described a system in which operators in North Dakota would use 4,500 MW of wind-derived electricity to power off-the-shelf electrolyzers. The system would then pressurize the hydrogen gas and feed it into 2-meter pipelines.

The economics of this scenario remains vexing. Its cost would be 30 to 45 percent more per unit of energy than that of building electrical transmission lines to link the Dakota wind farms with the power grid serving Chicago, the new study estimates. However, Leighty points out, breakeven could occur in other scenarios. For example, today's considerable research efforts could lead to fuel cells that are somewhat cheaper to make and operate.

Moreover, he adds, there are potential advantages to a hydrogen pipeline that economists currently find hard to value. For instance, it would—as its natural gas counterparts do—store several days' worth of energy in the system. Therefore, temporarily becalmed turbines wouldn't disrupt downstream operations.

Also, it may prove less expensive to add distributed sources, such as wind turbines, to a pipeline route than to a transmission line.

Finally, there's the potential that pollution taxes in the future could significantly increase the cost of fossil fuel systems and tilt the economic balance in favor of emissions-free power, including fuel cells. Indeed, many energy analysts argue that the only way the United States could ever meet the projected caps on carbon emissions being discussed under the Kyoto Protocol (SN: 12/20&27/97, p. 388) would be to tax people who spew carbon dioxide from combustion engines and boilers.

Within a decade, Leighty predicts, a technology harnessing wind to create hydrogen for fuel cells could become economically competitive.

—J.R.

For now, however, most developers aim to use wind to generate electricity for distribution and sale by commercial utilities. Already, a few big projects are in the works.

A 300-MW wind farm is being constructed along the Oregon-Washington border, where transmission lines can handle the load. This project will become the world's largest wind-harvesting system.

But a Goliath 10 times that size, tentatively named the Rolling Thunder project, is on the drawing board of Jim Dehlsen, founder of the pioneering wind-turbine company Zond, which was bought out by power giant Enron Corp. If built, this South Dakota network of turbines would be "one of the largest energy projects of any kind in the world," points out Brown of the Earth Policy Institute.

Windustry, however, is banking on small farm- and ranch-owned operations becoming the backbone of U.S. wind-power development. To foster that, Daniels says, her group is trying to see if next year's federal Farm Bill can include incentives for the development of farmer-generated commercial power. These might include guaranteeing bank loans, easing access to transmission systems, and facilitating development of wind-electric cooperatives. After all, Daniels argues, "wind is the best new crop to come along in many years."

Moreover, she points out that small-scale wind farming keeps much of its income in the local economy. That's good because wind resources are often strong in areas with poor soils. In such areas, it doesn't take a huge investment to make a big impact. A few Minnesota wind farms "have basically resurrected several small towns," Taylor notes.

And that's just the beginning, Brown says. He anticipates that people—call them wind prospectors—skilled at pinpointing the best places for wind farms could soon assume a role "comparable to that of the petroleum geologist in the old energy economy."

Many Switch to Green Power[3]

BY AMY WORDEN
PHILADELPHIA INQUIRER, JANUARY 22, 2001

In the coal region southeast of Pittsburgh, atop a mountain stripped bare by decades of mining, eight giant wind turbines hum softly around the clock. Erected last spring, the $10 million project is Pennsylvania's first large-scale venture in using the power of the wind to produce electricity.

These sculptural steel towers, high in the Allegheny Mountains, not only stand in stark contrast to the rusting coal-dredging equipment abandoned nearby. They represent a new and surprising reality: People are willing to pay extra for energy that doesn't pollute.

Green is the term for energy produced in part from naturally renewable resources, such as wind, water and the sun. The electricity created this way flows into your house through the same wall sockets—no new wiring (green or otherwise) required.

But it is considered cleaner than electricity produced by traditional fossil or nuclear fuels because it does not produce air pollution or create nuclear waste.

Since deregulation went into effect two years ago for Pennsylvania's 5.4 million residential customers, 473,852 have switched electricity providers. Of those, almost 20 percent have chosen companies that offer green energy.

Among New Jersey's 3.2 million residential customers, by comparison, only 69,000 have switched providers since the market opened there in August 1999, and only a fraction of those opted for green suppliers.

In Pennsylvania, the state Capitol, the state prison system, and the Rittenhouse Sheraton in Philadelphia are among the commercial customers now getting their electricity this way. Last summer, Langhorne, Bucks County, became the first town outside California to choose green energy.

The fact that more than 80,000 households are paying as much as several hundred dollars extra a year to help reduce toxic emissions from coal-fired plants is remarkable, say environmentalists and state officials.

"It's clearly significant that people are willing to spend more than lower-priced options in the marketplace for a product that pollutes less," said John Hanger, president and chief executive officer of PennFuture, an environmental advocacy group.

---

3. Article by Amy Worden from *Philadelphia Inquirer* January 22, 2001. Copyright © *Philadelphia Inquirer*. Reprinted with permission.

"I would have assumed people would look for the best deal or the best come-ons."

State consumer advocate Irwin Popowsky said: "I think it's one of the most positive developments in Pennsylvania. That people are willing to pay more, if there is the assurance that it comes from a cleaner resource, that's a tremendous development."

Of the roughly 80,000 households in the state that have switched to green, about 65,000 are in the Philadelphia area, according to PennFuture.

By the spring, as many as 115,000 of the area's two million households could be buying environmentally friendly electricity, which would make the city one of the nation's largest regional consumers of green energy.

In California, the only other deregulated state offering consumers a green alternative, 170,000 customers have signed up, but the extra costs are subsidized by the state.

Surveys show that 40 percent of consumers are . . . willing to pay more for cleaner power.

Ann Bartholomay, a Newtown resident and business owner, said she liked the idea of cleaner energy but decided to switch to Green Mountain Energy Co. only in December, when the cost became competitive with Peco Energy Co.'s.

"Economics was part of the reason I made the switch," she said. "I could have done better with savings, but I want to do good. I'm a '60s child. I think windmills are good."

She said she considered the extra $10 a month on her utility bill a donation to the environment.

PennFuture's Hanger points out that most customers select Green Mountain's Ecosmart plan, the company's least expensive of three options, priced according to the percentage of renewable resources used.

The Ecosmart plan uses only 1 percent renewable resources and is therefore not certified under "green-e," a national standard that requires 50 percent renewable resources (and a standard that some environmentalists say is too stringent).

Still, Hanger said, any movement away from coal and nuclear power helps the environment.

Green Mountain is the leading provider of green power in Pennsylvania, blending renewable resources such as wind and solar energy and gases from landfills with traditional nonrenewable resources such as natural gas.

The company operates the wind farm in Garrett, which produces enough energy to power 2,500 houses. Green Mountain also operates the state's largest solar-generation facility, in Conshohocken.

It is one of three suppliers offering green energy options in Peco territory.

Surveys show that 40 percent of consumers are concerned about the environment and are willing to pay more for cleaner power, Hanger said. For now, a Green Mountain customer in the Peco area with a $100-a-month electric bill, for instance, pays between $3 and $15 a month extra, depending on the plan he or she chooses.

In some parts of the state, the green option is still prohibitively expensive. In Pennsylvania Power & Light territory in central and northeast sections of the state, customers could pay up to $350 a year more.

Here at the Green Mountain Wind Farm, atop the state's second-highest peak, clean-energy advocates are banking on the growth of wind power as a viable alternative in a state that staked its claim on coal more than a century ago. t

"The potential for wind is tremendous," said John Holtz, Green Mountain's regional public affairs director.

Two other large-capacity wind farms are scheduled to open in western Pennsylvania this year, Holtz said.

"That green-market suppliers are developing green resources in Pennsylvania is very positive," Popowsky said.

And a hardscrabble community built on coal is embracing an idea once considered a pipe dream. Craft shops are offering miniature replicas of the turbines. Landowners want to lease their properties for wind farming. Seven thousand people have navigated the narrow winding roads to the remote site since the turbines began spinning last May.

Donald A. Decker, who with his brother, Robert, owns the two properties where the Green Mountain turbines stand, worked almost 50 years in the coal mines and lived amid the roar of machinery and fumes of coal dust on his property. Now he's a proud wind farmer with no backbreaking work to do. He gets paid to watch the wind blow.

"It's miserable," Decker said with a smile from the driveway of his tidy farmhouse. "I can sit here and watch the bladesurn. They relax you. I ought to have more of them."

Sunny Solution[4]

BY PETER FIMRITE
SAN FRANCISCO CHRONICLE, NOVEMBER 9, 2001

The only objects as impressive as the four Newfoundland dogs at Michael and Kim DeLongis' house in Petaluma are the four humongous batteries that help keep them comfortable. The 2,500-pound batteries, locked away in a cabinet in the garage, are part of one of the largest, and most ingenious, residential solar systems in California.

Two hundred and fourteen solar panels on the garage roof provide the electricity that charges the batteries that send electric currents through two inverters that power the house. It's enough to warm up two people and four pony-size dogs even on the coldest of days.

"The technology is so simple, it's crazy for people not to do it," said Michael DeLongis, 38, as he showed off the batteries and the whirring inverters that look like electric guitar amplifiers. "There are no moving parts, so nothing can break, and it's a clean source of energy."

Fed up with rolling blackouts, rising electricity rates and concerned about the future following PG&E's bankruptcy, consumers all over the state are turning to solar as a legitimate alternative. It is a phenomenon that has all the markings of a movement.

The DeLongis family's system is the second-largest residential solar power generator on the PG&E grid. Michael DeLongis, a contractor, installed the system this year because he was tired of paying huge electricity bills. The lifelong Republican also wanted independence from the powerful utilities.

"I have a well, I have a generator, and now I have solar," he said. "I'm a green Republican."

After three decades of false promises and wishful thinking by its proponents, solar is no longer just an eco-hippie pipe dream. Improvements in photovoltaic products and a 200 percent drop in prices over the past 30 years have combined with rising energy costs to drive the market.

San Francisco voters just approved revenue bonds that will, among other things, provide $100 million to install photovoltaic panels on the roofs of buildings. The mayor of San Diego wants to create "solar farms" to burn landfill gases. Even Texaco, one of solar's perceived enemies, is getting into the act, using an array of solar panels near Bakersfield to power the company's drilling operations.

---

4.   Article by Peter Fimrite from *San Francisco Chronicle* November 9, 2001. Copyright © *San Francisco Chronicle*. Reprinted with permission

The nationwide market for solar products is now worth $2 billion, according to the Energy Commission. Some federal and state energy officials are predicting it will grow to $10 billion by 2010 and double every three years for the next 20 years.

Locally, the residential market is taking off. PG&E currently has some 160 applications for residential solar systems that would hook into the grid. Each of the 31 homes in the Cherry Blossom development in Santa Cruz County

Solar is no longer just an eco-hippie pipe dream.

is being equipped with solar, and another 1,000 homes near Sacramento will be solar powered.

The technology has, in some cases, advanced beyond PG&E's ability to handle it. The largest system, Ken Adelman's 150-kilowatt system near Watsonville, is so powerful that PG&E took steps to limit the amount of energy he can feed into the grid and be compensated for, prompting a heated battle that may ultimately be decided in court.

"It's astronomical growth," said Peter Gregson, the owner of Advanced Solar, Hydro, Wind Power Co. in Mendocino County, which helped design the DeLongis system. "We've increased our business probably 50 percent since January. This industry is looking at PG&E's problems, the power crisis and deregulation as an opportunity."

Gregson said that until the power crisis, 95 percent of the systems he installed were in the boonies, mostly for people living "alternative lifestyles."

Now, people in urban and suburban areas are joining the sun power movement.

For the DeLongis family, tapping the sun's rays wasn't cheap by any stretch, costing $125,000—and a second mortgage on their home—to fully install. But the couple expects to get half the money back through a California Energy Commission rebate.

The commission's Emerging Renewables Buydown Program, which started in 1998, pays those who feed electricity back into the grid up to 50 percent of the cost of the system.

On most days the DeLongis system generates considerably more electricity than the couple can use. That's impressive, considering how much electricity flies around their cavernous two-story house.

Last December, before the solar system was installed, the couple's PG&E bill was $650. That included the cost of hundreds of Christmas lights and warming lamps for a litter of Newfy puppies.

But even on non-holidays, the DeLongis family is not one to conserve, having consistently run up electricity bills of $200, and that was before the power crisis.

Kim says it takes energy to keep the four big, lazy purebreds happy.

"They need baths. They need to be blow dried," she said, patting one of the dogs on its meaty head. "They are just big furry children."

The 214 solar panels produce close to 75 kilowatts of power a day, which is fed through the same kind of inverter used to power the space shuttles. The average home uses 24 kilowatts a day, according to energy officials.

The excess electricity is automatically fed into the PG&E grid. The DeLongises get credits for the electricity they feed back, allowing them to tap into PG&E for free if they ever run out of solar during a particularly long stretch of rain.

On one recent bill, their electricity costs came to $0.00, and the gas charges were $19, but a recently installed electric water heater and clothes dryer should eliminate even that. They reckon the system will pay for itself in money saved within 10 years. And if they ever decide to sell, their solar-powered house will be worth more.

"With solar, people realize they can be completely independent and eliminate the potential for a utility price increase," Gregson said. "They can go up on the top of a mountain and have a 6,000-square-foot house and a swimming pool using solar, or live in the city and never have to experience a blackout."

Soybeans and Corn Power Their Way into Fuel Options[5]

By Laurent Belsie
Christian Science Monitor, March 5, 2001

So you think America's too dependent on foreign oil? Some people say the time is ripe for bio-fuels:

- As early as May, St. Louis public-transit passengers could be riding buses powered by a mix of diesel and soybean oil.

- Coming off a banner year, the corn-based ethanol industry plans to expand production to record highs.

- Then there's Edgar Leightley's method. The Pennsylvania farmer got so tired of high oil prices he started burning corn to heat his home with a new furnace that's selling like hot cakes.

While burning food for fuel isn't new—Rudolph Diesel used peanut oil to power one of his engines a century ago—the idea has suddenly become a lot more practical. Fuel prices are so high and crop prices so low that politicians and fleet managers are taking a new look. Already a hit in Europe, bio-fuels are renewable, clean-burning, would stretch the U.S. fuel supply, and help stabilize its farm economy.

They even smell nice. Users report the truck exhaust from soybean-based bio-diesel smells like French fries.

Even if crop prices go back up, experts say bio-fuels could have a bright future if the U.S. makes it a high priority to curb global warming.

"There is no doubt that these renewable energy systems are carbon neutral. So you just grow it and burn it and things just recycle," says Don Erbach, national program leader of engineering and energy with the research arm of the U.S. Department of Agriculture (USDA). "If that really gets to be a serious issue and we really have the international will to address this issue, I could see where it would be a big thing."

Ever since gasohol hit the pumps in the 1970s—with mixed results—corn farmers have urged motorists to use more ethanol. Now it represents the third-largest use for corn after animal feed and exports.

Last year, the industry produced a record 1.6 billion gallons of ethanol, and it is expanding. Archer-Daniels-Midland in Decatur, Ill., which makes about half of America's ethanol, plans to boost its ethanol capacity by a fifth.

The biggest surprise this year is bio-diesel. Testing has suggested the diesel-soybean-oil mix works just as well as regular diesel (except in particularly cold weather) and burns much cleaner. But the standard mixture—80 percent diesel, 20 percent soybean oil—has proved far too expensive. Now, thanks in part to low soybean prices, prices have fallen from $4 a gallon to between $1.25 and $2.25, says the National Biodiesel Board, an industry trade group based in Jefferson City, Mo.

That's almost on par with regular diesel and close enough for many users to take a new look. The USDA has begun mixing bio-diesel with heating oil to heat a dozen buildings—and two dairy barns—at its Beltsville, Md., research facility. A Phoenix concrete company has converted its 100 trucks to run on 100 percent soybean fuel.

Testing has suggested the diesel-soybean-oil mix works just as well as regular diesel . . . and burns much cleaner.

"We would very much like to use it," says Lyle Howard, manager of product development for the Bi-State Development Agency, which runs the mass-transit system in metropolitan St. Louis. He's preparing a proposal to use bio-diesel in the agency's 580 buses and 63 para-transit vans later this spring.

Subsidies help. In November, the USDA set aside $150 million for each of the next two years to pay ethanol producers for increasing the use of bio-fuels, such as ethanol and bio-diesel. At least five states are now looking at enacting tax incentives to further encourage bio-diesel use.

As a result, production has gone through the roof—from 500,000 gallons in 1999 to 5 million gallons this year. The USDA projects its program alone will boost output another 36.5 million gallons.

Such numbers don't begin to rival the 56 billion gallons of regular diesel produced annually. But if the petroleum industry is forced to move to low-sulfur diesel by 2006, as the Bush administration announced last week, soybean oil could become a key lubricant that allows the fuel to work, proponents say.

Lubricants, in fact, represent another promising market for crops. Instead of petroleum products used to grease semitrailer couplings, railroad tracks, and chain saws, researchers have developed crop-based alternatives that are now just as cheap and more environmentally friendly. "We hope this year will really blow the top off the marketplace," says Lou Honary, director of the Ag-Based Industrial Lubricants Research Program at the University of Northern Iowa in Cedar Falls.

Such projects are helping to funnel money back into the nation's long-suffering farm economy. Ethanol production alone adds some $4.5 billion in farm revenue a year, according to the Renewable Fuels Association in Washington.

Still, some farmers are taking matters into their own hands. "I'm not selling corn to turn around and buy oil," says Mr. Leightley, a grain and vegetable farmer in Centre Hall, Pa. Instead, he's burned about 350 bushels of corn kernels to heat his farmhouse, saving half of what he would have spent using fuel oil. The company that made his furnace—Ja-Ran Enterprises in Lexington, Mich.—used to struggle to sell 10 of its biomass furnaces a year. Now, it's selling 24 a month.

"People are getting their gas bills and their propane bills," says owner Randy McLachlan, "and the stragglers are beginning to wake up."

Micropower: Wave of the Future[6]

BY LUCY CHUBB
ENVIRONMENTAL NEWS NETWORK, JULY 16, 2000

First National Technology Center, the central computer and processing facility for First National Bank of Omaha, does all the things that similar facilities do.

Mainframe computers and file servers whir away, keeping track of automated teller machine transactions and credit card purchases. Mechanical check sorters hum along, doing their thing.

What sets this Omaha, Nebraska, operation apart is its source of power. The electronic and mechanical equipment housed in the tech center doesn't run on juice from the grid. Its primary source of power is fuel cells that aren't connected to the centralized electricity network.

The bank turned to this technology in June 1999 after a costly computer crash in 1997.

"We needed a reliable source of power," said Brenda Dooley, president of First National Buildings, a sister company to First National Bank that oversees the tech center operation. "We do not want to be down because it affects our customers."

Dooley and co-workers weighed the options and chose a fuel-cell system from Sure Power Corporation in Danbury, Connecticut, which offers computer-grade electricity and 99.9999 percent reliability. The power grid offers 99.99 percent reliability—a significant difference over the long run.

"It's working very well," said Dooley. "It's meeting our expectations and we are very happy."

The decision to turn to an independent, off-the-grid power system is a trend that Seth Dunn hopes will catch on. In "Micropower: The Next Electrical Era," a report released today by the Worldwatch Institute, author Dunn asserts that the world's prevailing power generation systems are incompatible with demands of the coming century and that alternatives must be put in place.

"We're beginning the 21st century with a power system that cannot take our economy where it needs to go," said Dunn. "The kind of highly reliable power needed for today's economy can only be based on a new generation of micropower devices now coming on the market. These allow homes and businesses to produce their own electricity, with far less pollution."

Micropower devices include fuel cells, solar panels, wind power and micro-turbines that use natural gas.

---

6. Article by Lucy Chubb from *Environmental News Network* July 16, 2000. Copyright © *Environmental News Network*. Reprinted with permission.

In his report, Dunn focuses on three areas in making the argument in favor of micropower technology: reliability, the environment and the developing world.

As the economy becomes increasingly computerized and businesses become more dependent on high technology to get the job done, a rock-steady flow of electricity is crucial.

"There is a huge need for very reliable, high quality power," Dunn said.

Even small fluctuations in voltage in a practically imperceptible amount of time can cause catastrophes. Power interruptions cost the United States as much as $80 billion in lost data and productivity each year.

Micropower systems such as the fuel cells at First National Technology Center offer an excellent, decentralized alternative for dependable energy. They also "cut back on the footprint" of energy use on the environment, said Dunn.

Combustion-based electricity plants are one of the main sources of pollution around the world. Micropower solutions produce few harmful emissions.

"Micropower solutions may be most consequential in the developing world," Dunn said, "where 'power poverty' is becoming as economically and politically unsustainable as power outages are in richer nations."

Central power stations in developing countries are generally unreliable and severely pollute the environment, causing major health problems. Moreover, Dunn said, these centralized plants do not come close to meeting the needs of citizens. As a result, about 1.8 billion people living in more remote areas do not have access to grid-based electricity.

Instead, they rely on dirty forms of electricity generation, such as diesel generators and kerosene lamps.

"In these parts of the world, decentralized technologies have enormous potential to bring power to the people," Dunn said, "allowing the development of stand-alone village systems and doing away with need for expensive grid extension. And for a rapidly growing urban base, small-scale systems can substantially reduce the economic and environmental cost of electrical services."

The response by the utility industry to potential of micropower is mixed. Some companies are investing in these technologies to prepare for the future; others are simply digging in their heels. Dunn said some electricity companies are charging exorbitant disconnection fees for customers trying to remove themselves from the grid.

Regardless, he sees the potential for a dramatic shift to micropower that could take the world—and the electrical power industry—by storm.

"The current way the market is set up is very short term," he said. "The large-scale electricity model appears to be collapsing under its own economic and ecological weight."

Micropower Goes Macro[7]

BY CRAIG OFFMAN
WIRED, APRIL 2001

The Gist: Business people and homeowners alike are learning that generating their own electricity is cheaper—and more reliable—than buying it from centralized power plants. What Y2K couldn't accomplish, California's energy debacle did: Consumers are getting serious about catalytic fuel cells and microturbine generators that run cleanly on natural gas or propane.

False Alarm: Centralized power forever! Naaah. Political salves will provide short-term relief, but the conventional power grid ignores basic laws of physics: Sending electricity over hundreds of miles of wire is a losing proposition.

Exhibit A: Got a laptop? If so, you're familiar with the heat its power converter gives off as it goes from 117 volts of household current to the 12 volts your computer runs on. That heat is energy lost forever. Now visualize the big picture: Huge power plants—perhaps hundreds of miles from your home—burn natural gas to heat water into steam that spins generator turbines; in the process, they vent 40 to 60 percent of the flames' energy into the environment as waste heat. An additional 50 to 60 percent of what's left is lost on the way to your house, as transformers and wires heat up in resistance to the current. "It's a fundamentally flawed system," says inventor Dean Kamen, rumored to have created a super-efficient Stirling engine that could serve as a household generator. "At the time these plants were set up, most were based on natural resources available in the area. But there are better ways to generate more power locally now." All Temperatures Controlled, a California ventilation contractor, gave up the grid in favor of its own gas-powered generator. Michael Holzer, a director at ATC, says, "Brownouts and power surges were costing us $5,000 per hour. Now we have a monthly savings of about $800 on utilities."

Words to Live By: "Why not just buy natural gas for your home and generate the 12 volts you need locally? Some of the waste heat could go to your hot-water tank."—Brent Van Arsdell, president of Kansas-based American Stirling Company

On the Rise: Micropower is already popular with people who need an uninterrupted source of electricity. California-based Capstone Turbine sold $7.1 million worth of generators last year to business

---

customers, half of whom use the units as their sole supply of juice. "At first, micropower will complement central power," says Seth Dunn of Worldwatch, an environmental think tank in Washington, DC. "But as the technology's costs come down, it will be more logical for everybody to have these kinds of systems."

The Energy We Overlook[8]

BY ROBERT U. AYRES
WORLD WATCH, NOVEMBER/DECEMBER 2001

Carbon dioxide is not the only greenhouse gas (GHG). In fact, it accounts for only about half of the climate warming effect. But as a practical matter, the clearest path to reducing GHGs is to cut out carbon dioxide emissions from the burning of fossil fuels. There are three generic means of doing this: to get serious about conservation; to substitute other energy sources for fossil fuels; and to capture and sequester the CO_2 from fossil-fuel combustion. Let's take them in order of priority. By "priority," I mean the potential of each approach to produce gains large enough not just to meet Kyoto-scale goals, but essentially to "zero out" CO_2 emissions in the coming century.

Conservation. The energy industry routinely brushes this off as a non-starter, by suggesting that energy conservation means energy deprivation. This argument echoes the predictions of nuclear industry advocates of a generation ago, who told us that if we tried to rely on conservation and didn't embrace nuclear power, we would find ourselves "freezing in the dark" (a pro-nuclear bumper sticker from the early 1980s). We didn't, because a critical part of conservation is *efficiency*—getting the same mileage, or lighting, or heat, with less energy than before, and in the 1970s we began learning how to do that. Now, there are some simple measures for achieving greater "end-use" efficiency that could be introduced quickly and would cost very little—and in some cases would actually pay for themselves in a few months or years.

There are also prospects for making more radical, longterm, improvements in energy efficiency, both in its production and in every stage of its use.

Substitution. The nuclear industry is already licking its lips about making a comeback in public esteem. It is advertising heavily that it offers a "clean air" alternative to coal and oil. But even if the problems that have caused the current public disenchantment with this industry could be solved overnight, nuclear power plants take up to ten years to design, site, and build. Moreover, they are not cheaper than fossil fuel-fired power plants. And the nuclear industry is relatively mature, so the prospects for sharp cost reductions are dim at best.

Other non-carbon power sources are more promising. They include hydro-electricity (especially from small "low-head" dams), wind power, solar power, geothermal power, ocean currents, and solar power satellites. At present, all except small hydroelectric facilities would be more costly than big central power stations, at least if the social and environmental costs of fossil fuel combustion continue to be ignored. Large-scale deployment of standardized mass-produced wind or solar powered units

To rely much more on conservation does not mean "freezing in the dark."

that could bring unit costs down dramatically could take two or three decades. There is an inevitable lag as market size and production costs move in synchrony. Geothermal power and ocean currents are wild cards that could be helpful in some, but not most, locations. Solar satellites are, for the moment, a very long shot.

According to conventional wisdom, none of these options can individually substitute for a large percentage of the existing fossil fuel-based energy supply. Taken together, however, these options can have a major impact. Moreover, conventional wisdom may be too myopic. The main reason wind and solar power are not usually taken seriously as potential substitutes for coal and oil in satisfying the power needs of most countries is that they provide only intermittent supply. The wind does not always blow and the sun does not always shine. But solutions to the problems of intermittency are feasible, and some are already under development.

Sequestration. Not to be confused with the sequestering of carbon in trees, this technological sequestration would involve intercepting emissions before they can dissipate into the atmosphere, and locking the carbon up. It appears that this approach may be quite cost-effective for certain large-scale users—especially coal fired steam-electric power plants, which currently account for about a third of the carbon dioxide produced in the United States. It also appears that the captured gas can be utilized quite productively, especially in repressurizing (extending the productive life of) aging oil and gas fields.

These three generic approaches are listed in inverse order of attractiveness to the existing fossil fuel energy complex. The third approach is the one the established non-nuclear energy industries will prefer, and they will lobby hard to secure government funding and support for it, while undermining efforts to promote alternatives. Every U.S. politician from a district with an oil well, a coal mine, or a gas pipeline will be tempted to support this approach, but only after the political pressure to "do something" about climate warming has become irresistible. It hasn't yet.

What Offers the Largest Potential for Cutting Greenhouse Gas Emissions?

The answer, perhaps surprisingly, is conservation. To rely much more on conservation does not mean "freezing in the dark." Nor does it mean depending on voluntary individual choices to reduce domestic consumption, although that can't hurt. What it really means is to sharply increase the efficiency with which energy is used throughout the economy.

In the early 1970s, electricity demand in the United States had been rising for 20 years at about 8 percent per year. Some voices urged conservation then, as they do now. So the U.S. government promoted several "independent" (but noticeably orchestrated) studies of the potential for conservation savings. These studies all concluded that (1) conservation was a good idea but (2) the maximum potential savings were only around 15 to 20 percent, and (3) demand was going to keep on rising at the same rate as in the immediate past. In other words, this potential for conservation savings would be used up by two or three years of growth at 8 percent per year.

The implication of these studies was that many more new power plants and refineries would be needed before the end of the century. That view was reinforced by the propaganda of the nuclear industry in the 1980s, when it was fighting for survival in the wake of the reactor accident at Three Mile Island. Those projections turned out to be wrong. And a main reason was that the potential for conservation was grossly underestimated then—as it is now. Dick Cheney and George W. Bush, in making their energy policy proclamations of 2001, could well have lifted their projections from that decades-old nuclear industry propaganda.

One reason for underestimating conservation potential arose from a fundamental but very wide spread misunderstanding of the science of thermodynamics. Most engineers and economists assumed that energy was already being used quite efficiently. It was the prevailing view, based largely on a study carried out by the Livermore National Laboratory for the Joint Committee on Atomic Energy (JCAE) of the U.S. Congress, which concluded that the U.S. economy in 1970 was achieving an astonishing overall efficiency of 47.5 percent. Efficiency was defined as the ratio of "useful" energy (output) to "total" energy (input). The finding was offered with a straight face, so to speak, despite the fact that not a single energy conversion system in operation at the time could come close to matching that performance.

The deception lay in the assumption that most of the heat energy consumed by industry, and by residences and commercial establishments for space—heating, cooking, hot water, etc., was being used as efficiently as possible—that is, at an efficiency of 70 percent or more. This underlying assumption was built into virtually all official studies and recommendations. If the country's economy as a whole were already 47.5 percent efficient, there couldn't be much additional potential for conservation. In effect, the JCAE conclusion, echoed by others, blinded policymakers and investors to the possibility that further efficiency gains were possible.

The implicit definition of efficiency used in that official study in 1973 (and still accepted by most of the industry people who haven't had a good course in thermodynamics) is deceptive. The error was pointed out clearly by a summer study sponsored by the American Physical Society in 1975. But that study received much less publicity—and is much harder to read—than the official report to the Congress's Joint Committee and its numerous clones and progeny.

What was the error? By definition, a measure of efficiency must be a fraction—a number less than one. The denominator of the ratio, on the bottom, must be the energy actually used, by the economy as a whole, or by some industry, or factory, or appliance. The smaller numerator, on the top, should represent the *minimum physically possible* amount of energy required to achieve the same outcome, whether in tons of steel produced or hamburgers cooked. Only by this kind of comparison can we see how much room there is for future improvement.

That is not always the kind of ratio used, however. The gas industry likes to advertise that gas furnaces are about 80 percent efficient—by which it means that 80 percent of the heat produced goes into the house and only 20 percent goes up the flue. No doubt, this is an improvement over the 50 percent or so that leaked out of the house from coal-burning furnaces in the past. But that comparison is quite misleading, because it is fundamentally wasteful to use very high temperature heat for home heating. The important clue is that 70 degrees F., the temperature you probably want to keep your house at, *is almost exactly the temperature of the waste heat from a conventional steam-electric power plant.*

Why is that a clue? Let us suppose, for purposes of argument, that electric power plants need not be big. (If electronics can be miniaturized, why not electricity generation?) Suppose the fuel used to heat your house were used instead to produce electricity. The electricity could run the lights and all the appliances in your house, and maybe your neighbor's house, and the waste heat would still take care of the space heating. In other words, the heat would be a free byproduct. (This sort of system—known as "district heating"—is quite common in Europe, by the way. It can pose practical difficulties due to the distance between the power plant and the housing. But that is irrelevant to the question of what the true energy conservation *potential* is for heating buildings.) Moreover, conventional electric power plants are not as efficient as theoretically possible. For instance, the overall efficiency can be raised considerably by putting a gas turbine in front of the steam turbine, and using the hot gases from the gas turbine to heat the steam. This so-called "combined cycle" can achieve efficiencies as high as

60 percent—because some of the heat energy overlooked by the conventional energy accounting is now being used. And still, the waste heat at the end of the cascade is enough to heat your house.

In short, the *minimum physically possible* amount of energy needed to heat the house could theoretically be provided by a scheme that makes far better use of the same fuel. In other words, the true efficiency of the gas furnace is far lower than advertised. The APS summer study estimated that space heating in 1970 by means of an oil- or gas-fired furnace, and distributed by hot water or steam radiators, was only 6 percent efficient, in contrast to the 70 percent or so assumed by the JCAE study.

By this sort of test, the U.S. economy in the mid-1970s was nowhere near 47.5 percent efficient. In 1975, I tried (with a colleague) to revise the JCAE estimate of the energy efficiency of the U.S. economy, based on the principles explained by the APS summer study. Our first estimate, using 1968–73 data, came up with less than 2 percent. A later effort, using 1979 data and slightly more conservative assumptions, arrived at about 2.5 percent. The methodology of such calculations is arguable, especially in regard to deciding on what is the "minimum possible." But even with more traditional assumptions on that point, the real efficiency of the U.S. economy (and others around the world) is not more than a few percent at most. A very safe estimate would be less than 5 percent. What this means is that there is ample room to cut energy consumption, *without* cutting standards of living, for a long time to come.

> *There is ample room to cut energy consumption, without cutting standards of living, for a long time to come.*

Policies to Pump-Prime Efficiency

Some economists have advocated a carbon tax on fossil fuels, to reduce demand for such fuels and thereby cut CO_2 emissions. Model exercises have explored the possibility, and it turns out (as one might expect) that the outcome of this exercise depends upon what is done with the proceeds of the tax. The models tend to suggest that if the tax money is returned to consumers, or if it is spent by the government, employment and the growth rate will fall. On the other hand, if the tax funds are used to cut the existing social-security and value-added taxes on labor—thus reducing the effective cost of labor vis-a-vis energy—the macro-economic effects (on employment and growth) can be beneficial. Moreover, if a fraction of the proceeds is spent on energy research and development (R&D), there may be a double dividend—an economic spur to go along with the reduced carbon emissions.

In the 1970s, the U.S. government did help considerably by introducing the Corporate Average Fuel Economy (CAFE) standards for new automobiles and the Public Utility Regulation and Public Utility Regulatory Policies Act of 1978 (PURPA). The first of these laws simply mandated gradually better fuel economy for cars, leaving it up to the manufacturers to figure out how to achieve that result. The carmakers did so (kicking and screaming, to be sure) mostly by reducing vehicle size and weight and improving aerodynamics and tires. By 1988, U.S. automotive fuel economy was over twice as high as it had been in 1972, and roughly equal to the averages for Europe and Japan. The other law, PURPA, allowed utilities to make more profits for stockholders by producing power more efficiently. This was a sharp change from the earlier system in which a utility could increase its income only by building new capacity at a standard rate of return on investment fixed by regulators. (In those days, any savings from more efficient operation had to be passed on to consumers as lower prices. So why bother?)

The new law aimed to do something else as well: to force utilities to buy excess power from private producers at the utilities' marginal cost of production. Anyone with a small windmill or hydroelectric dam, or a cogeneration plant (making electricity from waste heat) could sell unneeded power back to the local utility at a known price. The utilities didn't like it, of course, because their control systems were not designed for decentralized production. They discouraged small suppliers as much as possible by imposing high "connection charges." Even so, many of them found themselves for the first time with excess capacity. A few utilities in regions that were growing rapidly and did *not* have excess capacity, notably in California, then found it worthwhile to help consumers save energy so as to delay the need for costly new generating capacity. (New plants were always more costly than older ones, due to rising land prices and increasingly strict environmental regulation.) This "demand-side management," as it was called by alternative-energy pioneer Amory Lovins, proved to be another effective tool for inducing conservation.

Lovins correctly predicted that thanks to such conservation, energy demand would not continue to increase at the historical rate. And indeed, between 1972 and 1988, conservation saved roughly a third of the energy that would have been needed had the industry and government experts been accurate in their forecasts.

Have we squeezed most of the potential from conservation by now? Far from it. In fact, we have barely scratched the surface. Take motor vehicles, the source of a third of the CO_2 emissions in the United States and other industrialized countries. The CAFE standards have not been tightened since the 1980s, and automotive fuel economy has ceased to improve. In fact, as most readers of this publication will be ruefully aware, the popularity of sports utility vehicles (SUVs)—classed as "trucks" to avoid strict regulation—has turned the trend in the wrong direction. Yet numerous studies confirm that fuel economy could be at least *tripled* by exploiting light but strong space-age composite materials (based on carbon fibers), using aerospace-related integrated design concepts that minimize the need for heavy steel frames, and using hybrid or fuel-cell propulsion units.

The Bush administration, under pressure from the auto industry, opposes any extension of CAFE standards. Yet, the U.S. auto industry did not suffer when the first CAFE law went into effect. (The period of increasingly intensive import competition from Japan in the 1980s was rough, for a while, but Japanese competition had much less to do with fuel economy than reliability.) A new CAFE standard, forcing fleet average fuel economy for all new vehicles (including SUVs) to twice the current level over a period of 15 years or so would be nearly painless for consumers and a very valuable inducement to technological innovation for the industry. And there is no reason to worry about the so-called "rebound effect" (the idea that increased efficiency will only lead to increased consumption) in this case. True, consumers might save some money by paying for less fuel, but alternative uses of that money would certainly be less energy-intensive than driving SUVs to the shopping mall.

By far the largest opportunities for increased end-use efficiency are to be found in the domain of residential and commercial buildings.

The impact of tougher CAFE standards on energy consumption could be magnified by another device, which Amory Lovins has called the "feebate." Each vehicle with a lower-than-standard fuel economy would be taxed in proportion to its excess use of fuel, while each vehicle with better than standard fuel economy would receive a proportional rebate. The money collected from the first group would pay for the rebates to the second group. This should satisfy the people who think that money collected by the government is always misspent.

The principles of the CAFE standards and the feebate scheme could just as well be applied to a number of other products, from refrigerators to home heating systems. In fact, by far the largest opportunities for increased end-use efficiency are to be found in the domain of residential and commercial buildings. Technologically, it is possible to reduce heating requirements for new buildings by as much as 90 percent, even in cold climates, by a combination of better insulation, better windows (triple-glazed with tight seals), and better design to utilize solar heat in the winter but exclude it in the summer. Extra costs for the insulation and high-performance windows are largely compensated for by the opportunity to meet all heating and cooling needs with a climate control system of lower capacity, thereby reducing capital costs. Solar photovoltaic (PV) panels on the roof can reduce the need for purchased power from the central utility. If the resulting structure costs a bit more—which it need not, in many locations—the difference would pay for itself in lower heating and cooling costs over its lifetime.

For older buildings, the potential is not nearly so great, at least on a per-building basis, but older buildings outnumber new ones. And retrofitting windows and roof panels can still cut heating needs substantially, while providing supplementary electric power at low marginal cost. Because a PV roof panel serves two functions (both power and heat), it spreads the cost base. In this application, PVs are already cost-effective in some locations and as manufacturing costs decline they will soon be more so.

Why aren't these things being done on a large scale already? Part of the answer is sheer inertia. The building industry is very decentralized. It is also very conservative, as are most of its customers. Very few builders have the technological expertise to exploit the most energy-efficient techniques, and most don't even know about the possibilities—mainly because there is no competitive pressure for them to investigate these things. The pressure on home builders is essentially to maximize usable (and visible) floor space per dollar. Consumers do not often ask about future operating costs. And more importantly, mortgage financing institutions do not insist that they do so, even though there is an apparent incentive for lenders to assure themselves that borrowers are aware of—and can afford to pay—the full costs of ownership, not just the mortgage repayments.

Public awareness campaigns, like those initiated in the 1970s in response to the oil crisis, can help. But the real gains from conservation, especially in the decentralized residential and consumer sectors, will probably require a push from regulation. There are a number of ways to do this, but the one I favor would be to work through mortgage lenders and utilities. New houses could be sold as a complete *package* of services, including the house and its equipment, all the necessary utilities (accompanied by long-term contracts), and insurance. The base cost to the buyer would include payments not only for the mortgage, but also for all primary utilities, on a pre-specified sliding scale of increased rates for increased energy use.

How could this be achieved? What's needed is an organization to do the "packaging" of the service components. As matters stand today, builders build; banks lend; utilities supply electricity or water or gas. Combining these services in a single consumer-friendly package could be beneficial for all concerned, and a clever entrepreneur should be able to make a profit by doing so. In a competitive free market, deregulated utilities should be prepared to compete for such business with the commercial builders (not the individual consumers), and the insurance companies should insure the delivery of promised utility services, as well as the usual protections against structural problems, accidents, and hazards.

Notwithstanding the conservatism of the residential housing sector, studies by the Lawrence Berkeley Laboratory and others indicate that with appropriate government policies in place, the residential and commercial use of energy can be cut by 50 percent or even more by mid-century, while the carbon emissions can be cut by 75 percent. Insulation, alone, will sharply reduce heat requirements. Most of the reduced amounts then required will be provided by solar collectors, small on-site generating units (fuel cells) with PVs, supplemented by electric heat pumps (as air-conditioning is now). The rest will come from natural gas or hydrogen. And the purchased electric power component can be made significantly less carbon-intensive by utilizing more low-head hydroelectricity, wind power, solar power, and industrial co-generation.

In those industries that have been traditional "big" users of energy (transportation, electric power generation, petroleum refining, metals, cement, glass, pulp and paper, and chemicals), fuels constitute such a large share of their overall costs that most of the obvious ways of cutting fuel use have been tried. In these industries, only more radical changes in technology will make much difference. In the electric power sector, the combined-cycle use of gas turbines and steam turbines has the potential to significantly increase thermal efficiency from its present plateau of 33 percent (including distribution losses) to 50 or even 60 percent. Newly constructed facilities already employ this technology. However, existing plants cannot be retrofitted for higher efficiency, so progress will occur slowly as total capacity increases and older plants are phased out and replaced. Unfortunately, the Bush administration seems intent on allowing old plants with obsolete pollution control systems to operate even longer, rather than encouraging their replacement.

In those "old economy" manufacturing and service industries in which energy is only a minor cost compared to those of labor and capital, the main incentives have been to save on labor, even if it means more mechanization and more use of energy. As a result, these sectors offer surprisingly numerous opportunities for what Amory Lovins has called "free lunches"—savings that cost nothing or that pay for themselves very quickly. Economists have been very skeptical about this, on the grounds that if there were major opportunities to save energy and money, the profit incentive would operate to make sure that such opportunities are not persistently overlooked. The fact remains that many such opportunities still exist, however—most likely because managers are much less concerned with finding small savings than they are with finding new markets and "growth."

Non-Carbon Alternatives

Of all the non-carbon alternatives[1]—water power, wind power, solar heat, photovoltaic electricity, tidal power, and geothermal power—the least expensive by far is low-head waterpower from small dams. In areas that have flowing streams and consistent rainfall, all that is needed to encourage more of these small generating units is to ensure that excess power can be fed back into the network efficiently. This is also a necessary—though not sufficient—condition for large-scale adoption of wind power and solar photovoltaic power.

Wind is the next cheapest. Western Europe is adopting wind power rapidly, albeit with some help from subsidies. But the subsidies have already created a significant market for the units, and this has brought a number of competitors into the field. Individual units are getting larger, now up to 1 megawatt, and costs are dropping sharply. A further 50 percent reduction is expected before 2010.

There is plenty of wind potential. Europe could generate two or even three times its current demand for electric power from wind. A similar potential exists in North America. However, so far the potential for stand-alone wind units has been relatively limited because of intermittence of both supply and demand. The grid has its *raison d'etre*. Until recently, it was technically quite difficult to integrate many small independent suppliers (that were also occasional net users) into the same grid that served the utilities' own large base-load plants. Most utilities have tried to discourage small retail power producers, either by means of high connection charges or legal restrictions. However, modern computer capabilities have largely eliminated the coordination problem, and it should not be a major barrier in the future.

Europe could generate two or even three times its current demand for electric power from wind.

For a long time, it has been assumed that local storage of energy—the thing that would really allow wind and solar power to take off—is not a serious option. But this is changing. The National Power PLC of Britain is developing a new type of regenerative fuel cell that is suitable for large-scale applications. As presently configured, it consists of tanks of sodium bromide and sodium polysulfide (there are many possible electrochemical couples). These concentrated salt solutions react electrochemically across a membrane, producing a cell voltage of 1.5 V. The cells can be combined in series (like any battery) to get higher voltages, and in parallel to get more power. The amount of energy stored is limited only by the size of the tanks. The storage capacity can be adjusted to a few minutes (to smooth out peaks) or to many hours. The technology has been tested in the laboratory and at pilot scale. A plant now being built in the United Kingdom is rated at 120 mWh of energy-storage capacity and up to 15 mW power rating.

It is true that at present, none of the alternatives mentioned (except hydropower) can compete economically with a large-scale conventional coal-fired power plant—assuming, as noted earlier, that the social and ecological costs of the coal-fired plant are ignored. However this does not mean that there are no cost-effective applications for renewables. On the contrary, there are already quite a few, with more to come. Mass production brings costs down dramatically—look at the history of the Ford Model T, or of the computer. Wind turbines and PV panels are still semi-customized products, made in rather small numbers. But in both of these industries, private investment is rising and capacity is growing at about 30 percent per year.

Getting CO_2 out of Circulation

Techniques for capturing and sequestering carbon dioxide from thermal power plants have been gaining in credibility (and funding) in recent years. The costs can apparently be kept quite low, provided there is a convenient way to store or otherwise dispose of the carbon dioxide. One possibility is deep sea disposal—dissolving the gas in sea-water at high pressure. However, the effect on ocean dynamics has not been modeled yet. Another possibility is to pump the carbon dioxide back into the ground, especially into old oil and gas fields. The heavier gas would tend to displace dissolved methane and thus increase the output of natural gas, at least slightly. Presumably, very little—if any—of the carbon dioxide would ever reappear at the surface. Both of these schemes would involve long-distance transportation on a rather large scale, most likely by pipeline or ship. The costs would be quite large for power plants far from an oil or gas field, or from deep water. A third possibility is to find a way of utilizing pure (or nearly pure) carbon dioxide in some useful and long-lived product, almost certainly a construction material of some sort. This idea seems far-fetched, but researchers at Los Alamos National Laboratory, among others, are working on it.

Toward a Hydrogen Economy

Throughout the stationary (non-transportation) world, energy delivered to the final point of use is increasingly in the form of electricity. This trend is expected to continue. However, storage difficulties do seem to restrict the long-term potential for use in vehicles. At present, there is no realistic substitute for liquid hydrocarbons as fuels for mobile power sources. Nevertheless, there is increasing interest in another possibility—hydrogen. For many years, hydrogen was thought to be too dangerous to use in vehicles, perhaps in part because of the famous Hindenburg fire back in 1937. However, cooler heads and recent research suggest that safe methods of storage, even in small quantities, are possible. Indeed, there may be an analogy with past experience with steam engines. Accidents were common (and often fatal) in the early days when the boiler was essentially a large teakettle, and when a leak could result in an explosion. The invention that made explosions impossible was the so-called monotube boiler—essentially a long tube inside a tank. There is no reason the same idea couldn't be applied to compressed hydrogen storage, perhaps on a microscopic scale (nanotubes).

Fuel cells have emerged, since the 1980s, as the great hope for a new generation of vehicular power plants. The big breakthrough was the development of a plastic membrane that allows protons (ionized hydrogen atoms) to pass through while blocking the passage of electrons.[2] This makes it possible to accumulate negatively charged electrons on one side of the membrane and positively charged protons on the other, creating a voltage difference. The availability of this material has enabled the development of the so-called proton-exchange-membrane (PEM) fuel cells, which are the basis of the most active current research programs, especially by Ballard Power Systems Inc. However, these cells also depend upon the availability of hydrogen as a fuel, and platinum as a catalyst. To be sure, the amount of platinum required has been cut dramatically since the first prototypes, but it cannot be eliminated alto-

---

Fuel cells have emerged, since the 1980s, as the great hope for a new generation of vehicular power plants.

---

gether. (Unfortunately, the less platinum is used per cell, the more uneconomic it is to recycle.) But platinum is an extremely scarce metal, and it is doubtful whether there is enough to support a whole new energy economy.

There are, however, a number of other types of fuel cells. Some are suitable for use in buildings, where both heat and power are needed. The most advanced is the phosphoric acid cell. Another possibility suitable for buildings is the high-temperature molten carbonate fuel cell. This can be paired with small gas turbines, using the waste heat. Composite efficiencies of 60 percent are possible. Several other types of cells are also attracting interest. And recent research has overcome some of the barriers that formerly seemed to block progress.[3]

Owning Your Share of CO_2

The usual policy prescriptions for cutting greenhouse gases include eliminating direct or indirect subsidies to fossil fuels, setting efficiency standards (such as CAFE), directly regulating emissions, and taxing emissions. Carbon taxes to reduce consumption are being seriously considered in the European Union.

I would like to propose something more radical: a system whereby every legal resident over some minimum age receives a fixed carbon entitlement, based on the annual quota for the country as determined by international negotiation along Kyoto lines.

This entitlement is equal for all legal residents over the minimum age, and it is recorded in a bank account. Parents of eligible children have control of their children's entitlements. The entitlements are tradeable, and have a market value. They can be bought and sold for money.

Here is how the scheme would work. Every time an individual makes a direct purchase of a carbonaceous fuel—say gasoline for the car—his or her carbon entitlement account is debited by the appropriate amount of carbon, as well as by the money price. The system works exactly the same way as a debit (money) card works today, except that there is a parallel unit of account in kilograms of carbon. Surplus individual entitlement units can be sold. The sale would be done through the bank where the account is held, at the prevailing market price, and the money value of the entitlement units sold would be deposited in the customer's money account at the same bank.

In this system, there are no automatic entitlements for businesses or other organizations. Unlike the widely discussed idea of CO_2 permit trading, it does not "grandfather" allocations to existing corporations; it requires those corporations to pay their way like everyone else. If a shop or a manufacturing firm needs to buy fuel it must use entitlements purchased in the market. This cost will, of course, add to the price of the goods or services being produced. The added cost is passed back to consumers, just as a carbon tax or a value-added tax is passed back. Thus, individuals pay indirectly for their carbon use, through higher prices for goods and services. Needless to say, energy-intensive services rise in price more than energy-conserving goods and services.

On average, people would receive a net income from the sale of surplus entitlements to finance the extra costs they have to pay. But low-income people with low levels of material consumption would gain from the scheme. Elderly stay-at-homes and bicycle-riding students would receive extra money from the sale of unused entitlements. They would receive more from the sale of surplus entitlements than they would have to pay in extra costs, amounting to a real income supplement. But for high-income people with high levels of material consumption—people with big display houses who drive SUVs towing large power boats, for instance—there would be an additional cost to buy the extra carbon units they consume.

The advantages of this scheme over a simple, uniform carbon tax are several. First, the level of the "tax" (i.e., the price of entitlements) would not be determined by politicians, but by the free market. Only the amount of carbon consumption (and emissions) to be allowed in that year would be fixed by government. Second, the scheme would explicitly recognize that rights to carbon consumption cannot be unlimited, and that as a limited resource they ought to be allocated equally to everyone. If politicians chose to award free consumption rights to certain favored users (say farmers) it would have to be done and defended openly. Third, and most important, the scheme would hit the wasters hardest and reward the conservers.

A Critical Consideration

A key point of this discussion is that economic growth in the long run depends upon continued technological innovation, not just marginal improvements in existing technologies. The potential for marginal improvements

> *Although giant companies dominate the economy, radical changes are rarely initiated by big firms.*

in any technology is always declining. For example, internal combustion engines improved dramatically in the half-century from 1876 to 1926. From 1926 to 1976, further gains were minor at best. Only radical changes can keep the forward momentum going. The fuel cell would be such a radical change.

But although giant companies dominate the economy, radical changes are rarely initiated by big firms. For example, the modern developments in fuel cell technology came from a small firm, Ballard Power Systems Inc., even though most of the basic research had previously been done by General Electric. Why didn't GE follow through? The short answer is that radical changes are disruptive, and big firms hate uncertainty. They are happy and profitable as long as things stay as they are. In a new ball game anything can happen. Today's profitable product can be an obsolete loser tomorrow. So, corporate giants are generally risk-averse. This does not prevent them from doing important basic research from time to time. Bell Laboratories invented the transistor, for example. But others, like the upstart Texas Instruments, exploited it. Hewlett-Packard and Xerox invented several important technologies underlying the PC. But Apple exploited them.

Small firms have different motivations. As long as things stay as they are, small firms remain small. In fact, they have trouble surviving against larger competitors. In order to survive and become more profitable they must get big, and that means taking market share away from some existing big firm, or—better yet—inventing a new market. Either way involves big risks. Most small firms that take big risks will fail. That is the way of the world. But when a risky venture succeeds, the payoff can be very big indeed. Most of today's big companies got that way by taking risks they would not take now.

The point I am making is that the Bush Administration is protecting the dinosaurs against the risktakers. But then, it is the dinosaurs who can afford to make big campaign contributions.

One last but most important point: those who resist any challenge to the fossil fuel economy claim that government intervention invariably inhibits economic growth. My claim is almost exactly opposite, and it is based on the real technological history of the past century: many of the most important technologies of our age came into being only *because* of government intervention. This is not to deny that government-operated services tend to be inefficient. But consider a few examples related to energy: the railroads across the western plains were built by private enterprise but only because of large land grants (subsidies) from the government. The U.S. interstate highway system was a tax-supported federal project. Large civil jet aircrafts depend on engines (and many other technologies) that were developed originally for military (government) purposes. The air traffic control system is still operated by the government. The big hydroelectric dams that provide both irrigation and electric power to several important areas were government-funded. Nuclear power was entirely based on reactor designs and fuels developed for nuclear weapons. In fact, the nuclear power industry still depends on a liability exemption mandated by congress. (You can't sue for damages in case of a nuclear accident.)

The Department of Energy supports extensive R&D programs in nuclear technology (including fusion power) and in several fossil-fuel related areas such as "clean coal." The government subsidizes Midwestern corn growers and grain processors to produce ethanol as a fuel. Why, then, is the U.S. administration so reluctant to support the really important new technologies like wind power, solar power and—above all, conservation—that could make the biggest difference?

Notes

1. Excepting the nuclear option, which is of highly questionable viability because of its unsolved problems of radioactive materials proliferation and disposal.
2. The idea of a filter that allows protons to pass while blocking electrons is somewhat contrary to intuition, because protons are much heavier than electrons. But because of the nature of nuclear forces, they are more tightly bound and thus smaller.
3. For more detail, see Worldwatch Paper 157, *Hydrogen Futures: Toward a Sustainable Energy System*, by Seth Dunn, published earlier this year.

III. The Holy Grails: Nuclear & Hydrogen

Editor's Introduction

N uclear and hydrogen are examined in this section together because both hold great promise as clean, renewable energy sources. In fact, some argue that they are the only realistic alternatives to fossil fuels. Today's nuclear power plants work through fission, splitting the atoms of a heavy element (uranium or plutonium) into lighter elements, in a controlled reaction that releases heat. (Nuclear fusion—the energy that powers the sun and stars—fuses hydrogen atoms together to release energy. Fusion has not yet proven practical for energy production, though the International Thermonuclear Experimental Reactor [ITER], a joint project between Russia, the U.S., Europe, and Japan, promises to be the first fusion device to produce more energy than it consumes.)

Nuclear power plants generate about 15 to 20 percent of the world's electricity. In France, 75 percent of all energy is nuclear. Nuclear power generation is highly efficient and produces few toxic emissions; for this reason, many say nuclear power is the best available response to global warming. Nuclear's detractors point to threats from terrorist attacks or reactor meltdowns, as well as the problem of radioactive waste, a byproduct of nuclear power generation that must be contained for at least 10,000 years. Bowing to such concerns, the German government has recently decided to phase out nuclear power.

In this section's first article, "Nuclear Energy Comes Full Circle," Charles Wardell visits Three Mile Island, the site of America's worst nuclear accident on March 28, 1979. Public support and government funding for nuclear energy dried up in its wake, and no new reactors have been built in the U.S. since. Wardell reports on a new type of pebble bed reactor that, by improving security and efficiency, may breathe new life into the nuclear industry. In an accompanying sidebar article entitled "Not So Fast," Edwin Lyman, scientific director of the Nuclear Control Institute, rebuts the claim that pebble bed reactors are meltdown-proof. Another sidebar article addresses some of the problems of containing radioactive waste in the planned site in Yucca Mountain, Nevada. (In early January 2002, after this article's publication, the U.S. Energy Department announced that it would recommend the use of Yucca Mountain as a containment site for nuclear waste.)

Christine Laurent investigates the claim that nuclear power is the best way to reduce the greenhouse gases that fuel global warming. Although the U.S. has pulled out of the Kyoto Protocol since Laurent's report, many other industrialized nations have agreed to reduce emissions and to help developing countries adopt clean energy technologies. Laurent writes that, while nuclear energy is one way to do this, many also advocate for reducing energy consumption—a challenge to Western nations' energy-intensive development patterns.

In "Managing a Nuclear Transition," Sam Nunn, a former senator and chairman of the board for the Center for Strategic and International Studies, argues that nuclear power represents the only way to meet rising energy needs while reducing carbon dioxide emissions. After having led a study of the global handling of nuclear materials, Nunn recommends several specific ways to ensure the safe use of nuclear energy.

Hydrogen is a tantalizing energy source: it is the world's most abundant element and it emits mainly pristine steam when burned. Many argue that hydrogen is the key to a renewable energy economy. Hydrogen gas can be used to store energy derived from the wind or the sun, for example, or to power a device known as a fuel cell, in which hydrogen and oxygen combine in a controlled reaction to produce electricity and heat. Some predict that fuel cells will be a disruptive technology, supplanting conventional batteries and the internal combustion engine and radically altering the way we live. ("It's hard to dismiss a technology that promises a way to kiss the sheikhs goodbye," David Stipp quipped about fuel cells, in *Fortune*.) Fuel cells are already in use in military applications and space-based equipment. The first commercial fuel cell products—including cars and buses—are expected to hit markets by 2003, although the costs for such technology will remain prohibitively high for most consumers.

In the first article about hydrogen, William Stevens reports on the 150-year trend away from the use of dirty solid fuels with high carbon content (wood and coal) toward lower-carbon fuels, such as natural gas and hydrogen. Though the energy we burn today is less carbon intensive, the world's burgeoning population and increased usage per capita mean more energy consumption overall. Nonetheless, de-carbonization represents the greatest hope for slowing the effects of global warming.

Seth Dunn describes projects underway in Hawaii, the South Pacific, and Iceland to create hydrogen-based energy systems. Such systems would take advantage of the islands' abundant geothermal, solar, and hydroelectric resources to make hydrogen energy, an attractive alternative to the high costs of importing oil. Dunn argues that the U.S. government must take a leadership role in launching the hydrogen economy by funding research, educating the public, and building the infrastructure that will be needed for a hydrogen age.

Nuclear Energy Comes Full Circle[1]

By Charles Wardell
POPULAR SCIENCE, August 2001

This is Ground Zero, where support for nuclear energy in America died 22 years ago. It's completely silent, save for the echo of my own voice. Above, a hawk hovers, riding the air currents created by the structure's unusual shape. The building is probably 100 feet in diameter, and would normally have 4 feet of water at its base.

Welcome to Three Mile Island's number two reactor, just outside of Harrisburg, Pennsylvania. More specifically, I'm standing in the middle of the now-empty cooling tower for the reactor, which, because of a faulty valve, nearly melted down on March 28, 1979. The reactor itself was housed in an adjacent building that's since been cleared. It's sealed too, because traces of radiation are still detected inside. A short distance away, clouds of water vapor rise from the plant's still-working reactor. It sounds more like a tourist attraction—nearby Bushkill Falls perhaps—as water cascades down the interior of the tower, cooled by air.

No one was killed or injured, but Three Mile Island—the event has assumed the plant's name—was America's worst nuclear accident. Plans for new nukes were shelved in its wake, and none has been built since. Yet this is the place, the most unlikely place in the United States, where nuclear power could soon be reborn.

And it's not just because Chicago based Exelon Corp., which runs Three Mile Island, is pushing for it. No, with electricity costs rising, politicians and the public alike are showing sudden support for nuclear energy. It has the full support of the Bush administration, which has made it a cornerstone of its recently unveiled energy policy.

But to truly understand the renewed buzz for nuclear, you have to travel 400 miles northeast, to the Massachusetts Institute of Technology in Cambridge. Here, Andrew Kadak, professor of nuclear engineering, holds two billiard-size balls that many believe represent the future of nuclear energy. The balls are the "pebbles" in something called a pebble bed modular reactor, a new type of plant that proponents say is safer and more efficient than current plants. It could even crank out electricity for less than a gas-fired plant, savings that would presumably be passed on to you. More important, considering our anxiety toward nuclear energy, it's immune to meltdowns. The technology could be implemented, possibly at Three Mile Island, within five years.

When Kadak, formerly vice president of the American Nuclear Society, came to MIT in 1997, nuclear power seemed doomed. So in January 1998, he challenged 11 students to design "a politically correct reactor" that would win acceptance from regulators and the public while giving gas a run for its energy-generating money.

All existing U.S. commercial reactors are "light water" reactors. They're powered by half-inch-diameter cylindrical pellets of uranium—like cutoffs from a 1/2-inch dowel—stacked up in 14-foot-long metal rods. Hundreds of rods are lowered into a water-filled reactor core. The uranium atoms give off neutrons, some of which crash into other uranium atoms, splitting them, generating heat, and knocking free more atom-splitting neutrons—the process known as fission. The water in the core carries the heat away to drive an electric turbine.

Kadak's students rejected light-water technology for this reason: If the coolant leaks away, the core heats up enough to melt. Instead,

---

While fission heats the pebbles to as much as 1,100°C, the coatings trap all radioactivity inside.

---

they found something they considered safer: a pebble bed research reactor that had run for 22 years in Germany, ("until Chernobyl came along and Germany got out of nuclear," Kadak says). It relied on fission too, but was fueled by eight-ball-sized pebbles, and rather than water coolant, it used helium gas.

The main safety feature is the fuel itself. Each pebble consists of roughly 10,000 "microspheres" of uranium dioxide the size of a pencil point. Each is in turn coated with several layers of graphite, and a silicon carbide outer shell. While fission heats the pebbles to as much as 1,100°C, the coatings trap all radioactivity inside. Once the fuel is spent, the coatings isolate radioactive decay particles for a million years—four times longer than it takes them to completely decay. Of course, they still need a permanent burial place (see "Yuck, a Mountain").

With the pebble bed, a Three Mile Island-type event couldn't happen, Kadak says. Even if the helium coolant completely leaked out of the core, the fuel wouldn't get hotter than 1,600°C, well below the 3,000°C or so needed to melt uranium. Plus the graphite coating is a great heat absorber.

A commercial pebble bed would produce 110 megawatts of electricity—one-tenth that of a large, light-water plant. Its core would consist, of a giant, upside-down bottle, 3½ meters in diameter and 8 meters high, filled with more than 400,000 pebbles. Pneumatic tubes would pull pebbles out of the spout at its base. They would be

Yuck, a Mountain

Even if pebble bed reactors prove invulnerable to meltdown, they don't solve a problem that's been around since the advent of nuclear power: what to do with the highly radioactive spent fuel that remains after making electricity. This waste has been piling up at more than 100 nuclear power plants around the country.

Today's nuclear reactors are powered by rods filled with uranium pellets, which remain hot—literally—for years after they are removed from a reactor. To prevent the rods from melting, they're stored in water-filled pools that also act as radiation shields. Problem is, the pools at many power plants are filled to capacity.

In 1982, the utilities operating nuclear plants began paying the federal government one-tenth of a cent for every kilowatt-hour of energy generated, on the condition that it begin accepting used nuclear fuel by 1998. It now appears that the earliest this could happen is 2010.

The government has looked at only one site seriously: Nevada's Yucca Mountain. Despite a barrage of studies, opponents say it has not been proven that waste buried deep within Yucca would remain immune to water seepage, earthquakes, and other potential hazards for 10,000 years—long enough for the waste to decay to levels equivalent to the natural radioactivity of uranium ore.

Some nuclear power plants have taken matters into their own hands, constructing massive concrete casks that store spent fuel aboveground. Meanwhile, the Democrats say they will not support plans to bury waste at Yucca. And since so many plants have had their licenses renewed, the Nevada site is no longer large enough to hold all of the nation's radioactive waste.

—Dawn Stover

continuously scanned, put back in the top if still usable, and sent to a sealed container if spent. All of this would happen automatically.

Kadak's students weren't alone in their fascination with the pebble bed. A few months into the project, Kadak learned that a South African electric utility called Eskom was doing similar research. Before he knew it, Eskom and Exelon had partnered to create a next-generation nuclear reactor. The MIT team, meanwhile, has since received more than $1 million in government funding to investigate fuels, reactor core physics, safety, and waste issues. It hopes to soon build its own research reactor at the Idaho National Engineering & Environmental Laboratory in Idaho Falls. Eskom and Exelon plan to build a working prototype in South Africa by the middle of next year. Exelon itself has invested more than $7 million and could submit a licensing application to the U.S. Nuclear Regulatory Commission by next summer as well. If all goes smoothly—a big if, indeed—Exelon could have a commercial pebble bed, operating in the United States by 2006.

Along with safety and efficiency, there is another major benefit to the pebble bed one that could make it easier to bring to America: Adding 110 megawatts at a time to existing plants would stir up less opposition than building a new plant. And because the reactor would be built with replaceable modules, it would cost 30 percent less per megawatt. In fact, it's more accurate to talk about assembly than construction. "You could take the balance of the plant, put it on a flatbed truck, and ship it to the site," he says. "That's a true innovation."

It would also cost less to run. Continuous fuel cycling eliminates the need for refueling shutdowns, which happen every 18 months at light-water plants. And the plant's gas turbines would be simpler and more efficient than the steam equipment used in conventional designs. So what? That means lower, more stable electric bills for you; a pebble bed could make power for well less than 2 cents per kilowatt-hour, versus 4 cents for a gas-fired plant (90 percent of all currently proposed new plants use gas).

Two decades after nuclear died . . . it's trying to rise from its own ashes, like a mythical phoenix riding the air currents above reactor number two.

But there's a catch. To meet those cost targets, Exelon will ask the government to license the technology without an emergency cooling system and without the airtight containment dome used in light-water plants. That will prove controversial. "Opponents keep raising the issue of containment," Kadak says. "But if one particle in a fuel ball fails, the radiation released is minuscule." He admits that the fuel itself will need regulatory scrutiny because the pebbles rely on their coatings for containment. And although there's no containment dome, the reactors themselves will be housed in a "citadel" strong enough to withstand the impact of a 747.

Exelon is more circumspect. Oliver Kingsley, chief nuclear officer, says the technology is far from a done deal. "It's a venture in the early exploratory stage," he says. "We're still doing design feasibility studies; we won't finish until this fall."

The pebble bed's fate also depends on that of the Price-Anderson Act, which limits a utility's liability for accidents. It expires next summer. Bush's energy proposal recommends its renewal, but if Congress declines, don't expect to see any new plants, let alone radically new designs. Also, the way the act is written now, a 110-megawatt plant faces the same liability as a 1,100-megawatt facility. Exelon will lobby to change that.

Kingsley says his company will only go ahead with the project "if the technology is deemed ready for commercialization, and if the economics prove to be competitive against other forms of generation." In other words, though the pebble bed looks promising, and its technology sound, its ultimate fate awaits the outcome of a just-beginning political and scientific debate.

Nobody knows which way the winds will ultimately blow. But one thing's for sure: Two decades after nuclear died, burning itself to death on the banks of the Susquehanna River, it's trying to rise from its own ashes, like a mythical phoenix riding the air currents above reactor number two.

Not So Fast

Edwin Lyman speaks in a soft, even tone. But his words are incendiary. The physicist, scientific director of the nonprofit Nuclear Control Institute in Washington, D.C., isn't worried about a Three Mile Island-type accident with a pebble bed modular reactor, the supposedly meltdown-proof technology that proponents believe will reinvigorate nuclear energy. His visions are more catastrophic.

Lyman is most concerned about the reactor's lack of a containment dome—instead it relies on the pebbles' graphite coating to prevent radiation leaks. "If the reactor remains intact, then their argument that it needs no containment is valid," he says. But a crack in the reactor itself could spell trouble. "You need to keep air away from the fuel, because graphite reacts with air and burns. That's what happened at Chernobyl." (MIT's Andrew Kadak, who helped develop the technology, calls the analogy preposterous, adding that in tests graphite has proven extremely difficult to burn.)

Lyman also points out that the technology would create 10 times the volume of waste as a conventional reactor. The radioactivity per kilogram would be lower, but it would still need to be transported and disposed of. And as for the modular construction system: "If you have a generic safety flaw, you have a whole fleet with the same defect."

Of course, not all threats are internal. To the outsider, the security at a nuclear power plant is impressive. To get into Three Mile Island's control room, for example, I was scanned for explosives and metal. The fence surrounding the facility is topped with razor wire. And the control room itself is surrounded by 3-foot-thick concrete walls. The entire facility is patrolled 24/7 by guards with assault weapons. Yet in 1993, an intruder drove a car into the front door of the turbine building and disappeared for 4 hours.

To test security, the federal Nuclear Regulatory Commission (NRC) periodically conducts mock attacks. And though utilities spend several months and as much as $1 million preparing, they fail at a rate approaching 50 percent. Failure, in this case, means that "adversaries reach the target," says David Orrik, who oversees the program.

Despite this track record, NRC quietly killed the program in 1998, then quickly reinstated it after public backlash. "In some cases, mock terrorists reached their targets without being challenged by a single response officer," wrote Massachusetts Congressman Ed Markey. Particularly troubling was a 1998 exercise at the Vermont Yankee plant near Burlington, where attackers reached the reactor itself. More recently, NRC has launched a pilot program in which utilities design their own exercises. "It's as if half the class failed the exam," says Orrik, "so the professor is letting them grade themselves."

But can the industry police itself? Soon after Orrik was quoted in the press about failure rates, he found himself under internal investigation. "My phone was tapped. They were into my files," he recalls. According to Jeff Duncan, Markey's legislative director, Orrik's experience isn't isolated. "The NRC has a bad track record with people who report problems," he says. "They've been the subject of investigation or harassment." Duncan knows of employees who have confidentially reported violations to the NRC, only to have the NRC report them to their employers.

Kadak, meanwhile, thinks the security question is moot anyway, pointing out that an attack on a pebble bed reactor would cause less damage than one directed at a light-water plant. "You've got 400,000 balls," he says. "What are you going to do, pick one up and throw it at someone?" Maybe. But at the very least, a successful attack would likely turn public sentiment against nuclear.

Since a pebble bed would be added to an existing plant, Lyman wants utilities to demonstrate that it won't spread security too thin. "There needs to be more public oversight," he says. "Any attempt to railroad these plants through will aggravate public opposition."

—Charles Wardell

Beating Global Warming with Nuclear Power?[2]

By Christine Laurent
UNESCO COURIER, February, 2001

For several years, the nuclear energy industry has attempted to cloak itself in different ecological robes. Its credo: nuclear energy is a formidable asset in the battle against global warming because it emits very small amounts of greenhouse gases. This stance, first presented in the late 1980s when the extent of the phenomenon was still the subject of controversy, is now at the heart of policy debates over how to avoid droughts, downpours and floods.

Today, few would deny the existence of global warming and the fact that we have to do something about it. The latest report by the Intergovernmental Panel on Climate Change (IPCC), released in November 2000, revised its own predictions upwards, forecasting that temperatures will rise between 1.5 and 4.5°C by the end of the century.

Human activity obviously shares part of the blame for global warming, starting in wealthy countries where industrial development began in the mid-19th century. It is estimated that in these countries, carbon dioxide (CO_2) and methane emissions increased by 70 percent and 145 percent respectively in 150 years. Gathered in Kyoto (Japan) in 1997, representatives of 38 industrialized countries signed a protocol under which they agreed to cut their overall CO_2 emissions by 5.2 percent of the 1990 level by 2012. The European Union is to account for 8 percent of the cuts, the United States 7 percent, Japan and Canada 6 percent. Russia was let off the hook because its economy collapsed after 1990. The effort should not be scorned. Studies show that if the trends of the past decade continue, these countries will emit 20 percent more CO_2 instead of 5.2 percent less by 2012.

But what's the best way of reaching these goals? The solutions range from using more nuclear power, developing renewable energy sources, encouraging mass-transit, implementing stricter energy-saving measures and planting forests. "There are all kinds of ways to cut CO_2 emissions," says Antoine Bonduelle, founder of the Climate Action Network, which includes 320 non-governmental organizations worldwide. "The negotiations become extremely complex if

---

2. Reprinted from the *UNESCO Courier*, February 2001, Christine Laurent.

all of these options are discussed at once." So much so that during a conference in The Hague in November 2000, the 180 countries failed to reach a consensus on how to apply the Kyoto protocol.

Improve Living Standards Without Suffocating the Planet

The main focus in the battle to cut greenhouse gas emissions is to reduce energy consumption, which accounts for approximately 80 percent of the discharges. The primary culprits are coal and oil burned by electrical power plants, heating installations and cars. The battle is all the fiercer because in the developing countries— which emit an average of 0.4 tonnes of carbon per inhabitant per year compared with three tonnes in the OECD—growth requirements will lead to a rapid increase in emissions. A recent report by Georges Charpak, winner of the 1992 Nobel prize for physics, warns that "if China pursues its current rate of industrial growth, by 2050 it will emit eight times more carbon dioxide than the entire industrialized world does today."

Seeking to avoid a situation where badly needed improvements in living standards go hand-in-hand with rising greenhouse gas emissions, rich countries agreed at the Kyoto summit to transfer more environmentally friendly and efficient technology.

In this context, nuclear power seems like a simple, efficient solution that is all the more attractive because it does not call into question our societies' energy intensive development patterns. The world's seven richest countries, which include the nations with the highest number of nuclear power plants, advocated this solution way back in 1989 at a G7 summit. "We recognize that nuclear energy plays an important role in cutting greenhouse gases," they proclaimed.

After the Kyoto summit, the European Union published in 1999 a study called "Dilemma," a title that speaks for itself. European experts sketched three different scenarios on CO_2 reduction, with nuclear power accounting for varying shares of total energy output. They concluded that the EU has no chance of delivering on its promise unless Europe boosts its nuclear power capacity by the equivalent of 80 percent of its current capacity. The European Atomic Forum, a nuclear industry trade group, claims that "in a single year, nuclear power has helped to avoid emitting 1.8 billion tonnes of CO_2 in the world, which, for Europe, is equivalent to the emissions of 200 million cars." Meanwhile, the OECD's Nuclear Energy Agency published a report in 1999 based on the hypothesis of tripling nuclear capacity by 2050. If that were to happen, CO_2 emissions would drop by 6.3 billion tonnes a year according to its calculations.

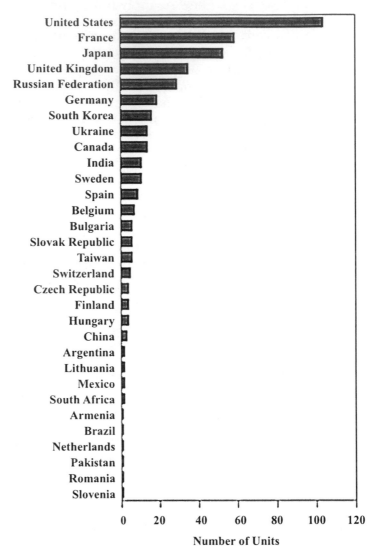

Operating Nuclear Power Plants Worldwide, 1999

Source: International Atomic Energy Agency, *Nuclear Power Reactors in the World 1999* (Vienna, Austria, April 2000).

At first sight, the figure sounds convincing enough to sweep aside any protest from the anti-nuclear camp. But emissions from human sources are increasing by 22 billion tonnes a year and despite planned restrictions in developed countries, they are projected to rise by 60 percent globally between now and 2020! This would bring them to 35 billion tonnes. While an annual reduction of 6.3 billion tonnes 30 years down the road should not be shunned, it's not enough to convince anti-nuclear advocates that this source of energy is the miracle solution to global warming.

Beyond expert circles, scenarios giving nuclear energy a new lease on life are running into mounting opposition from society at large, shaken by the 1986 Chernobyl catastrophe and growing concern over how to manage nuclear waste. Almost all developed countries have put their nuclear programmes on hold. In 2000, not a single reactor was under construction, on order or planned in North America or Western Europe. Today, the staunchest advocates of nuclear power are in Asia—Japan, Korea and, to a lesser extent, China and India—and in Eastern Europe. At the end of 1999, the Vienna-based International Atomic Energy Agency counted 38 reactors in the planning or construction stages in 14 countries. Many of them are unlikely to be commissioned because of the economic slump in the countries that have ordered them.

In 2000, not a single reactor was under construction, on order or planned in North America or Western Europe.

Anti-nuclear advocates also argue that the abundance of energy allowed by nuclear power plants discourages energy-saving measures. Furthermore, investing important sums in nuclear energy means earmarking less for exploring alternative energy sources. A report by the World Wildlife Foundation notes that nuclear energy does not necessarily lead to a reduction in CO_2 emissions. "The United States accounts for 5 percent of the world's population, over 25 percent of CO_2 emissions and 29.4 percent of nuclear-generated electricity. Conversely, China accounts for 21.5, 13.5 and 0.6 percent, respectively. A closer look at the trends between 1980 and 1997 reveals that the penetration of non-fossil power sources [especially hydraulic] in China during this period allowed that country to cut back on only 10 million tonnes of carbon, compared with 430 million tonnes had Beijing adopted energy-efficiency measures."

The World Energy Council also argues that the best way to reduce consumption is to improve energy efficiency. The potential is tremendous. Estimates suggest that savings could reach 30 to 50 percent in Europe, and 30 to 70 percent in the United States. This thinking, advocated by non-governmental organizations for a long time, is gradually winning support from political leaders, including in France (which obtains 75 percent of its electricity from nuclear plants), where the government has just adopted a national plan to improve energy efficiency.

"For three years, we've been working on translating the political commitments made at Kyoto into specific measures," said French Environment Minister Dominique Voynet after The Hague conference, where she led the European delegation. "To achieve that goal, consumer patterns in the rich countries will have to change and more efficient power sources will have to be offered to developing countries. But it will be difficult."

Co-generation is a new technology being adopted in the United States and Europe, and could show the way. A coal or oil-fired electric power plant is made to produce heat and electricity at the same time, recuperating up to 90 percent of the energy produced instead of an average 45 percent with conventional plants. Combined-cycle technology (a gas turbine combined with a steam turbine) has made tremendous strides and offers an energy efficiency rate of 60 percent, while remaining cheaper than a classic fuel-burning plant.

Individual consumers can also make a difference. Lighting manufacturers have calculated that the use of low-consumption bulbs each year saves electricity equivalent to the output of several

---

A few tenths of a kilowatt-hour of savings on a refrigerator represents, on a worldwide scale, the output of several dozen nuclear power plants.

---

nuclear reactors. In the same vein, household appliance and electronics manufacturers are designing products that guzzle less energy. A few tenths of a kilowatt-hour of savings on a refrigerator represents, on a worldwide scale, the output of several dozen nuclear power plants.

Lastly, in addition to energy-saving appliances, research suggests that renewable energy sources such as hydrogen, solar and wind power are bound to acquire a new dimension in the 21st century. Today, Germany's windmills generate as much electricity as four nuclear power plants. According to the Danish BTM-Consult, windmills could soon supply over 10 percent of the world's electricity—half the current nuclear output. The debate is far from cooling.

Finally, the anti-nuclear camp has taken another feather out of its cap. Mycle Schneider, head of the World Information Service on Energy (WISE), points out that the nuclear sector's CO_2 emissions are far from negligible when the whole production chain is looked at, not just the plants themselves. This includes construction, extraction, treatment, conversion, transport and the stocking of waste. Taking this into account, WISE calculates that for France, the nuclear chain's share of total emissions ranges from 1.6 to 9.1 percent. More importantly, Schneider stresses that nuclear plants only produce electricity, whereas our societies also have significant heating needs. Emissions by plants that produce both electricity and heat would be higher than those obtained with more efficient gas plants . . . The nuclear industry's response is being anxiously awaited.

Managing a Nuclear Transition[3]

By Sam Nunn
Washington Quarterly, Winter 2000

The future of our planet rests on the safe handling of nuclear power and materials. Nuclear energy is a God-given resource, and its proper management is vitally important today and for future generations. However, our record to date in forecasting its future development is questionable.

Sir Ernest Rutherford, one of the greatest contributors to the world's understanding of atomic structure, stood before the British Association for the Advancement of Science in 1933 and declared, "The energy produced by the breaking down of the atom is a very poor kind of thing. Anyone who expects a source of power from the transformation of these atoms is talking moonshine."

In 1958, Lewis Strauss, the second chairman of the U.S. Atomic Energy Commission (AEC), the forerunner of the Nuclear Regulatory Commission (NRC), asserted that nuclear power would make electricity so cheap that it wouldn't be worthwhile to meter it. And in 1971, the AEC predicted that by 1985 worldwide nuclear-generating capacity would reach 570,000 megawatts, more than half of which would be in the United States. These predictions, though in radically different directions, all look pretty silly today.

Our record over the years at managing and utilizing nuclear technology for peaceful purposes has been much better than our record at forecasting its future. But that record also leaves much to be desired, as evidenced by Three Mile Island and Chernobyl.

Notwithstanding this history of questionable predictions, today there are some 430 nuclear units operating around the globe that are safely and efficiently providing more than 16 percent of the world's electricity.

The long-term need for nuclear energy in the world's overall energy mix is clear. Indeed, we are witnessing a confluence of trends that makes a very strong case for the continued and future need for nuclear energy.

The first is demographics. In this decade alone, one billion people will join the ranks of humankind. Within little more than a generation, projections indicate that an additional four billion will be added to the planet's population.

The second trend is economic growth, arising not only from these demographics, but also from the fundamental desire on the part of every society on earth to advance their standard of living. Almost two billion of the six billion people on the planet today do not have access to electricity. In addition, there is a huge disparity in the energy consumption levels, and therefore the quality of life, between industrialized and developing countries.

The third trend—the world's demand for energy—results from the first two, and it is expected to double over the next 25 years alone. To meet this increased demand, the World Energy Council predicts that more than three thousand giga-watts of new electricity generation—that is, three million mega-watts—which is equal to the current worldwide installed capacity, will be required.

> *The world's demand for energy . . . is expected to double over the next 25 years alone.*

Based on the noted historical predictions, unless there are unexpected miraculous breakthroughs in some alternative energy sources, the great preponderance of new energy required over the next 25 years will come from coal, natural gas and nuclear power.

The fourth trend is the emerging imperative to address the threat of climate change or environmental damage. The International Atomic Energy Agency (IAEA) estimates that global carbon dioxide emissions will increase by 50 percent by the year 2010 because of the growth trends.

Widespread concern over climate change is not new. It extends back well before the Kyoto Convention to the Rio Conference of 1992, and indeed earlier to the Toronto G-7 Summit of 1988. But what is new is the emerging scientific and political consensus that global warming is a reality and the growing realization that greenhouse gases may have serious economic, environmental, and social consequences. And what also is new is the willingness to pursue global, binding agreements among nations to protect our environment.

Nuclear Power Challenges

These four trends will bring about a renewed, intensified examination of nuclear electric power as a source of clean energy supplies. However, there are several key challenges that face the nuclear power industry and the incorporated nuclear technology, if this new opportunity is to be grasped successfully. First, there is an absolute requirement for the highest levels of safety. Second, nuclear power must be economically competitive. Third, the weakest links in the global nuclear chain must be strengthened. This third challenge includes the highly contentious issue of nuclear waste management.

The global nuclear power industry has no alternative but to ensure a consistently high level of safety. The public only grudgingly tolerates nuclear energy today, does not fully understand its advantages, and not surprisingly finds it easy to overdramatize the risks. But one thing is certain, the public will not tolerate the use of nuclear reactors to generate electricity if it believes the price of that electricity will be the prospect of an accident at some unforeseen time.

Improved safety and reliability of nuclear plants across the globe depend first and foremost on the utilities and the plant owners and operators. National regulators must be vigilant in carrying out their responsibilities as must the IAEA and the World Association of Nuclear Operations (WANO), which must work together in a cooperative effort to ensure the highest levels of operational safety. All 32 countries operating nuclear electric generating plants worldwide are members of WANO. This achievement certainly allows for more international cooperation and communication between the plants.

Hans Blix, former director general of IAEA, succinctly linked public acceptance and nuclear safety in a speech at the Uranium Institute in 1997. He said, "Probably only a prolonged, relatively problem-free operation of nuclear power plants will dispel misgivings about the use of nuclear power." The challenge then becomes how to make and keep nuclear electric power economically competitive.

A deregulated energy marketplace will place enormous pressure on nuclear plants to produce electricity that is competitive on a cost basis with coal, oil, and natural gas. Our citizens will be the losers if that economic pressure causes nuclear operators to take short cuts with nuclear safety—or if competition erodes free and open exchange of safety information.

There is, however, evidence that it is possible to actually improve upon safety levels in a highly competitive environment. The airline industry, for example, has enjoyed a steadily improving safety record since deregulation of its operations began twenty years ago. There is little or no reason to challenge that nuclear electric power cannot emulate this success.

Within the nuclear industry it has been demonstrated that the safest plants are also the most efficient. High levels of safety and good economics are not mutually exclusive. Indeed, there is now clear evidence that good economic performance and high levels of safety are not just complementary, but codependent.

The Nuclear Waste Issue

The global nuclear chain includes many additional links, some of which today are fragile. These fragile links include the efforts to avoid the spread of nuclear weapons material and the acquisition of nuclear weapons by terrorists, whether by individuals or by rogue nations.

Our nuclear world has changed dramatically following the dissolution of the Soviet Union in 1991. During the Cold War, the two nuclear superpowers confronted each other all around the globe. It was a world of very high risk, but also of high stability, mainly because of the grave danger of escalation in nuclear weapons. Today, the world has really flipped upside down. The danger of nuclear war is very low, but also stability is very low. It's as though our house has come through an earthquake unscathed, but now is in danger of being eaten away by termites.

For the last eighteen months, I have led a Center for Strategic and International Studies panel consisting of U.S. and international experts charged with examining the subject of global handling of nuclear materials. Five issues-related panels were established, including experts from the IAEA, Sweden, Great Britain, the People's Republic of China, Japan, Argentina, the United States, and Russia. A coordinating committee summarized the findings and recommendations of the five task forces, and the summary was then submitted to a senior policy panel for review and approval.

Our vision of Global Nuclear Materials Management is of a world in which all nuclear materials are safe, secure, and accounted for, from "cradle to grave," with sufficient transparency to assure the world that this is the case. To accomplish this awesome task, a number of steps are essential and several are urgent, as follows:

- The United States must reengage and reinvigorate its leadership in the nuclear arena by stronger support of existing plants, particularly through licensing renewal by promoting research and development in the peaceful use of nuclear technology and by supporting the infrastructure for training and retention of nuclear engineers.

- Resolution of the long-running conflict between the U.S. government and U.S. nuclear utilities over responsibility for and approaches to spent fuel management, which now appears to be in prospect, is essential if the seriously damaged U.S. leadership is to be repaired. The Nuclear Energy Institute deserves great credit for the recent progress being made on this issue.

- As the President's Committee of Advisors on Science and Technology recently recommended, the United States should undertake a new international cooperative initiative to promote safe and proliferation-resistant interim storage of spent fuel, in both national and international facilities. The leading nuclear nations should seriously explore the establishment of an international facility for storage and disposal of spent nuclear fuel, possibly in Russia, possibly in the United States, or elsewhere. Russian atomic energy minister Yevgeny Adamov, for example, has proposed to the Russian Cabinet of Ministers that Russian environmental law be changed to permit storage of nuclear spent fuel from any country. Wherever the location, there are many obstacles to be overcome, but the world must give serious consideration to this concept.

- The United States and the international community must launch a new program of fissile-material threat-reduction which should focus on efforts to buy, consolidate, secure, monitor, and reduce weapons-usable nuclear material stockpiles and ensure sustainable nuclear security for the future. Several separate but interconnected initiatives are required to carry out this step.

The U.S. and Russian governments must place top priority on the successful implementation of the U.S.-Russian highly enriched uranium (HEU) purchase agreement which calls for the purchase of 500 tons of HEU from dismantled Russian weapons over twenty years.

A new comprehensive program of further HEU and plutonium purchases should be launched by the United States and other nations for the purpose of converting thousands of tons worth of weapons materials to peaceful reactor fuel. This would increase transparency and provide hundreds of millions of dollars annually to help Russia stabilize its nuclear complex. Additional purchases should be a top priority, with the proceeds to be directed to an auditable fund to help Russia pay for and improve its safe and secure handling of nuclear, chemical, and biological materials and weapons. These purchases should include not only HEU from dismantled weapons, but also those small, vulnerable HEU stockpiles located at small research facilities.

> *We must accelerate and expand U.S.-Russian cooperative efforts to consolidate nuclear material in Russia at fewer locations.*

We must accelerate and expand U.S.-Russian cooperative efforts to consolidate nuclear material in Russia at fewer locations. The effort underway to improve security, accounting, and consolidation of nuclear materials merits international support and should not be resource constrained. The physical devastation that could be caused by one weapon made from stolen materials and used against one target in a populated area would dwarf in dollar costs, let alone the cost in human lives, the cost of this program.

The vast stockpiles of plutonium and HEU built up over the decades must be converted to forms much less usable in nuclear weapons. As former Russian minister of atomic energy Victor Mikhailov once said, "Real disarmament is possible only if the accumulated huge stocks of weapons-grade uranium and plutonium are destroyed." As arms reductions proceed, these stockpiles should be reduced in parallel to roughly equivalent levels in the United States and Russia, suitable to support whatever agreed warhead levels remain, but not large enough to permit a rapid return to Cold War levels of armament. Importantly, this effort should be supported by the leading nuclear nations—not just the United States.

Commercializing the Russian excess defense infrastructure is clearly a priority but must be preceded by defense conversion. Commercialization is complicated by an undeveloped market economy and by a relatively immobile labor force. Development of a commercial environment conducive and safe for foreign investment is a critical issue that can only be resolved through a mutually advantageous relationship between the United States public and private sectors and the Russian government.

To continue with the list,

- The nuclear insecurity problems in the former Soviet Union can only be successfully addressed with a true spirit of partnership among Russia and other former Soviet states, the United States, and other leading nuclear nations. The United States and the international community must cooperate with Russia in its efforts to reduce the size of its nuclear complex through dismantling, conversion, or commercialization of facilities.

- The necessary effort to improve security for U.S. nuclear secrets in the aftermath of charges of nuclear espionage must not impose constraints that would undermine the ability of U.S. personnel to cooperate with Russia and other countries to ensure that nuclear materials do not fall under hostile control.

- Finally, the IAEA plays an absolutely central role in safeguarding nuclear material and working with states to improve its management worldwide. Having been limited to a zero-real-growth budget since the mid-1980s, the IAEA urgently needs additional funding, and the United States and other major nuclear powers should redouble their efforts to provide it.

World Safety and Security

Nothing could be more central to international security than ensuring that the essential ingredients of nuclear weapons do not fall into the hands of terrorists or proliferating states. Effective controls over nuclear warheads and the nuclear materials needed to make them are essential to the future of the entire global effort to reduce nuclear arms and stem their spread.

At the same time, the protection of public health and the environment in the management of all nuclear material—from nuclear weapons to energy production—remains a critical priority. The management of both safety and security worldwide will be essential to maintaining nuclear fission as an expandable option for supplying the world's greenhouse-constrained energy needs in the twenty-first century.

These steps cannot be taken by any one nation acting alone. World safety and world security are at stake, and world funding must be mobilized in response. The fundamental requirement is leadership. The time to act is now before a catastrophe occurs.

Global Economy Slowly Cuts Use of Fuels Rich in Carbon[4]

BY WILLIAM K. STEVENS
NEW YORK TIMES, OCTOBER 31, 1999

Even as the world's expanding population and economy increase atmospheric concentrations of carbon dioxide that scientists say are warming the earth, the global energy system is moving steadily away from the carbon-rich fuels whose combustion produces the gas.

Experts say atmospheric levels of carbon dioxide may be double that of the pre-industrial era by the end of the next century. But they also say the levels would be much higher except for a trend toward lower-carbon fuels that has been going on for more than 100 years, but has been largely unnoticed except by a small band of energy specialists.

The question now, they say, is whether the trend can be accelerated enough to stave off or lessen what many scientists believe is a potentially disruptive global warming.

For nearly a century and a half, fuels with high amounts of carbon have progressively been replaced by those containing less. First wood, which is high in carbon, was eclipsed in the late 19th century by coal, which contains less. Then oil, with a lower carbon content still, dethroned King Coal in the 1960s.

Now analysts say that natural gas, lighter still in carbon, may be entering its heyday, and that the day of hydrogen—providing a fuel with no carbon at all, by definition—may at last be about to dawn.

As a result, the experts estimate, the world's economy today burns less than two-thirds as much carbon per unit of energy produced as it did in 1860. In the United States, they estimate, the trend toward lower-carbon fuels combined with greater energy efficiency has, since 1950, reduced by about half the amount of carbon spewed out for each unit of economic production.

But because economic growth and population growth have been so rapid over the decades, overall atmospheric concentrations of carbon dioxide have steadily risen, to the point that the concentrations may well have doubled by the year 2100.

Mainstream scientists say that this much carbon dioxide could warm the earth, on average, by 3 to 5 degrees Fahrenheit. By comparison, that is about half as much as it has warmed since the depths of the last ice age 18,000 to 20,000 years ago.

---

4. Article by William K. Stevens from *New York Times* October 31, 1999. Copyright © *New York Times*. Reprinted with permission.

A change of this magnitude would likely have widespread consequences for the world's climate, weather and human life.

Now, as representatives of 150 governments meet in Bonn in the latest round of global talks on measures to further reduce carbon-dioxide emissions, analysts both in and out of industry say that the next quarter-century is shaping up as a period of technological and economic ferment offering a chance to accelerate the trend toward a low-carbon economy and, eventually, a no-carbon one.

In Bonn, the delegates are trying to work out the details of an agreement forged two years ago in Kyoto, Japan, that could speed up the trend. Their work is not expected to be finished for at least a year, and the Kyoto agreement still must be ratified by a sufficient number of countries after that.

However that may turn out, "the decarbonization of the energy system is the single most important fact to emerge from the last 20 years of analysis" of the system, said Dr. Jesse H. Ausubel, an expert on energy and climate at Rockefeller University in New York City. Dr. Ausubel predicts that this evolution will produce a carbon-free energy system by the end of the 21st century.

Among some recent signs of the trend are these:

- The Federal Energy Information Administration reported last week that emissions of carbon dioxide by the United States had increased by an average of 1.37 percent a year in the 1990s— only about half the 2.6-percent rate of growth in economic production. Analysts say the discrepancy is evidence that the economy is being decoupled from carbon.

- The agency reported this month that the same is generally true in China, the biggest consumer and producer of coal in the world, where coal production has been reported to be dropping lately. "China has dispelled a commonly held notion that economic growth and energy consumption are necessarily coupled," the report said.

- In December, Honda will introduce in the United States a high-efficiency, low-emissions automobile powered partly by gasoline and partly by self-generated electricity. It is said to run at 60 miles per gallon of gasoline in town, and 71 on the highway, and to travel 600 to 700 miles on a tank of gas.

Toyota has introduced a similar "hybrid" automobile in Japan, and these cars are "literally kick-the-tires examples of the decarbonized economy," said Hal Harvey, president of the Energy Foundation, a partnership of foundations that promotes energy efficiency and renewable energy.

Other auto makers are also planning hybrids, which are being viewed as a transition, ultimately, to vehicles powered by hydrogen fuel cells that emit no carbon. In its planning, the General Motors

Corporation has "embraced fuel cells as the technology of choice," but with hybrids coming first, said John Williams, the leader of the company's internal team on global climate issues.

And while auto companies are looking down that track, some of the world's biggest energy companies are looking to provide the appropriate fuels. Hydrogen, in particular, has attracted fresh interest.

Until recently, "the hydrogen option was seen as rather distant," said Ged R. Davis, an executive of Shell International in London who analyzes such questions for Royal Dutch/Shell, one of the world's largest energy companies. "Now it is looking closer, perhaps over the next decade or two," Mr. Davis added. "Most of the energy and car companies are looking at this rather seriously." Shell itself has established a hydrogen subsidiary.

In the nearer term, hydrogen would be used in fuel cells for cars, trucks and industrial plants, just as it already provides power for

---

Hydrogen would be used in fuel cells for cars, trucks and industrial plants, just as it already provides power for orbiting spacecraft.

---

orbiting spacecraft. But ultimately, hydrogen could also provide a general carbon-free fuel.

The world energy system will not change overnight, of course, if it changes at all. And new products must ultimately stand the test of the marketplace. But some analysts say that the next two decades or so will be a time of unusual pressure for change, both for environmental and economic reasons, in which companies will be driven to compete for survival and dominance in some sort of emerging new energy system.

Whether companies are seriously pursuing new options or merely preserving them for the future, experts say there seems little doubt that the long-term trend toward decarbonization is real, and that it will most likely continue even in the absence of any shift to hydrogen or renewable energy sources like wind and solar power.

"The future decarbonization rate is likely to be at least as high as the historical one" of about three-tenths of a percent a year, said Dr. Nebojsa Nakicenovic, an expert on energy and the environment with the International Institute for Applied Systems Analysis, a research group in Laxenburg, Austria. The institute was one of the first groups to study the question.

Oil accounts for the biggest share of global energy consumption today, followed by coal and, closely, by natural gas. In most of the world except the United States and China, said Dr. Ausubel of Rockefeller University, coal is either defunct or on the way out, and natural gas will increasingly displace it.

According to several recent analyses, Dr. Nakicenovic said, recoverable natural gas now appears far more abundant than had been previously thought. The burning of gas produces, on average, only about a third of the carbon dioxide per unit of energy of coal, and about two-thirds that of oil.

Gas not only can fuel fixed facilities like industrial plants and furnaces, it can also be processed to produce hydrogen for use in carbon-free fuel cells to power automobiles and generate electricity. In those cells, there is no combustion; instead, hydrogen reacts chemically with oxygen to produce electricity. But when hydrogen is extracted from gas, the residual carbon must somehow be disposed of, possibly by pumping it back into depleted oil and gas wells.

Dr. Ausubel predicts that natural gas will become the dominant fuel of the next 40 to 50 years. If so, that alone would be enough to continue the long-term decarbonization trend.

China, which some experts think will emerge as the biggest carbon-dioxide emitter of the 21st century, has greatly reduced its energy consumption per unit of economic output, has closed several coal mines, is seeking to modernize industrial and power plants and is moving toward natural gas, many analysts say.

Not least, they say, the Chinese are worried about the health effects of coal's air pollution. Nevertheless, the Energy Information Administration reported last week, China's coal demand is expected to double by 2020.

So while the trend toward a carbon-free economy may continue, Dr. Ausubel says, it might not move rapidly enough to assuage the fears of those who are most concerned about global warming. He says that if the trend continues to evolve more or less naturally, with business as usual, it will take another century or so to decarbonize the energy system fully.

By then, he predicts, atmospheric concentrations of carbon dioxide will be around 500 parts per million, nearly double what they were before the industrial revolution. Mainstream scientists say that would be enough to change the earth's climate substantially, make droughts, heat waves and floods worse and raise the sea level to heights that would threaten many low-lying coastal areas and islands. Some analysts say that 500 parts per million is a best-case estimate, and that business-as-usual could cause a tripling of pre-industrial carbon-dioxide levels.

Other experts think that concentrations could be held substantially below 500 parts per million if the trend toward decarbonization were to accelerate. Mr. Harvey of the Energy Foundation says "prospects are excellent" for an acceleration.

And Mr. Davis, the Shell executive, says his company's analyses suggest that if the proper incentives were in place, new energy technologies could be adopted broadly enough to bring about a peak in oil use and carbon-dioxide emissions by about 2020. After that, there would be a decline.

One sort of incentive might lie in the Kyoto agreement, which calls for a group of 39 industrialized countries to reduce their carbon dioxide emissions by an average of 5 percent below 1990 levels over the period 2008 to 2012. One mechanism for doing this is a system whereby a country that exceeds its reductions target can earn money by selling that extra reduction to another country that is having trouble meeting its target. A similar system, involving company-to-company trading, has been proposed for the United States.

While negotiators struggle over the terms of such arrangements and politicians wrangle over putting the Kyoto accord into effect, many energy analysts seem to agree on one thing: The ultimate goal ought to be a carbon-free economy based largely on hydrogen. Dr. Ausubel, for one, predicts that such an economy will materialize.

Many would agree with Mr. Williams of General Motors: "I think I'm on pretty solid ground in saying the long-term vision is hydrogen. But there's a lot of work between here and there."

Hydrogen as the Way Toward Sustainability[5]

BY SETH DUNN
THE HUMANIST, NOVEMBER/DECEMBER 2001

Hermina Morita has a grand vision for Hawaii's energy future. A state representative, Morita chairs a legislative committee to reduce Hawaii's dependence on oil, which accounts for 88 percent of its energy and is mainly imported on tankers from Asia and Alaska. In April 2001, the committee approved a $200,000 "jumpstart" grant to support a public/private partnership in hydrogen research and development, tapping the island state's plentiful geothermal, solar, and wind resources to split water and produce hydrogen for use in fuel cells to power buses and cars, homes and businesses, and military and fishing fleets.

The grant grew out of a consultant study suggesting that hydrogen could become widely cost-effective in Hawaii this decade. The University of Hawaii, meanwhile, has received $2 million from the U.S. Department of Defense for a fuel cell project. Possibilities include Hawaii becoming a mid-Pacific refueling point, shipping its own hydrogen to Oceania, other states, and Japan. Instead of importing energy, Morita told a San Francisco reporter, "Ultimately what we want . . . is to be capable of producing more hydrogen than we need, so we can send the excess to California."

Leaders of the tiny South Pacific island of Vanuatu have similar aspirations. In September 2000, President John Bani appealed to international donors and energy experts to help prepare a feasibility study for developing a hydrogen-based renewable energy economy. The economically depressed and climatically vulnerable island, which spends nearly as much money on petroleum-based products as it receives from all of its exports, hopes to become 100 percent renewable-energy-based by 2020. Like Hawaii, it has abundant geothermal and solar energy, which can be used to make hydrogen. And like Hawaii, it hopes to become an exporter, providing energy to neighboring islands. "As part of the hydrogen power and renewable energy initiative we will strive to provide electricity to every village in Vanuatu," the government announced in its October 5, 2000, issue of *Environment News Service*.

Hawaii and Vanuatu are following the lead of yet another island, Iceland, which amazed the world in 1999 when it announced its intention to become the world's first hydrogen society. Iceland, which spent $185 million—a quarter of its trade deficit—on oil imports in 2000, has joined forces with Shell Hydrogen, Daimler Chrysler, and Norsk Hydro in a multimillion dollar initiative to convert the island's buses, cars, and boats to hydrogen and fuel cells over the next thirty to forty years. The brainchild of a chemist named Bragi Arnason and nicknamed "Professor Hydrogen," the project will begin in the capital of Reykjavik, with the city's bus fleet drawing on hydrogen from a nearby fertilizer plant and later refilling from a station that produces hydrogen on site from abundant supplies of geothermal and hydroelectric energy—which furnish 99 percent of Iceland's power. If the project is successful, the island hopes to become a "Kuwait of the North," exporting hydrogen to Europe and other countries. "Iceland is already a world leader in using renewable energy," announced Thorsteinn Sigfusson, chair of the venture, in March 2001, adding that the bus project "is the first important step toward becoming the world's first hydrogen economy."

Jules Verne would be pleased— though not surprised— to see his vision of a planet powered by hydrogen unfolding in this way.

Jules Verne would be pleased—though not surprised— to see his vision of a planet powered by hydrogen unfolding in this way. After all, it was in an 1874 book entitled *The Mysterious Island* that Verne first sketched a world in which water—and the hydrogen that, along with oxygen, composed it—would be "the coal of the future." A century and a quarter later, the idea of using hydrogen—the simplest, lightest, and most abundant element in the universe—as a primary form of energy is beginning to move from the pages of science fiction and into speeches of industry executives. "Greenery, innovation, and market forces are shaping the future of our industry and propelling us inexorably toward hydrogen energy," Texaco executive Frank Ingriselli explained in April 2001 to members of the Science Committee of the U.S. House of Representatives. "Those who don't pursue it, will rue it."

Indeed, several converging forces explain this renewed interest in hydrogen. Technological advances and the advent of greater competition in the energy industry are part of the equation. But equally important motivations for exploring hydrogen are the energy-related problems of energy security, air pollution, and climate change—problems that are collectively calling into question the fundamental sustainability of the current energy system. These factors reveal why islands, stationed on the front lines of vulnerability to high oil prices and climate change, are in the vanguard of the hydrogen transition.

Yet Iceland and other nations represent just the bare beginning in terms of the changes that lie ahead in the energy world. The commercial implications of a transition to hydrogen as the world's major energy currency will be staggering, putting the $2 trillion energy industry through its greatest tumult since the early days of Stan-

dard Oil and Rockefeller. Over 100 companies are aiming to commercialize fuel cells for a broad range of applications—from cell
phones, laptop computers, and soda machines to homes, offices,
and factories to vehicles of all kinds. Hydrogen is also being
researched for direct use in cars and planes. Fuel and auto companies are spending between $500 million and $1 billion annually on
hydrogen. Leading energy suppliers are creating hydrogen divisions, while major carmakers are pouring billions of dollars into a
race to put the first fuel cell vehicles on the market between 2003
and 2005. In California, twenty-three auto, fuel, and fuel cell companies and seven government agencies are partnering to fuel and
test drive seventy cars and buses over the next few years. Hydrogen and fuel cell companies have captured the attention of venture
capitalist firms and investment banks anxious to get into the hot
new space known as ET, or energy technology.

The geopolitical implications of hydrogen are enormous as well.
Coal fueled the eighteenth- and nineteenth-century
rise of Great Britain and modern Germany; in the
twentieth century, oil laid the foundation for the
United States' unprecedented economic and military
power. Today's U.S. superpower status, in turn, may
eventually be eclipsed by countries that harness hydrogen as aggressively as the United States tapped oil a
century ago. Countries that focus their efforts on producing oil until the resource is gone will be left behind
in the rush for tomorrow's prize. As Don Huberts, chief
executive officer of Shell Hydrogen, has noted: "The
Stone Age did not end because we ran out of stones,
and the oil age will not end because we run out of oil."
Access to geographically concentrated petroleum has
also influenced world wars, the 1991 Persian Gulf War,
and relations between and among Western economies,
the Middle East, and the developing world. Shifting to
the plentiful, more dispersed hydrogen could alter the
power balances among energy-producing and energy-
consuming nations, possibly turning today's importers
into tomorrow's exporters.

Countries that focus their efforts on producing oil until the resource is gone will be left behind in the rush for tomorrow's prize.

The most important consequence of a hydrogen economy may be
the replacement of the twentieth-century "hydrocarbon society"
with something far better. Twentieth-century humans used ten
times as much energy as their ancestors had in the thousand years
preceding 1900. This increase was enabled primarily by fossil
fuels, which account for 90 percent of energy worldwide. Global
energy consumption is projected to rise by close to 60 percent over
the next twenty years. Use of coal and oil are projected to increase
by approximately 30 and 40 percent, respectively.

Most of the future growth in energy is expected to take place in
transportation, where motorization continues to rise and where
petroleum is the dominant fuel, accounting for 95 percent of the

total. Failure to develop alternatives to oil would heighten growing reliance on oil imports, raising the risk of political and military conflict and economic disruption. In industrial nations, the share of imports in overall oil demand would rise from roughly 56 percent today to 72 percent by 2010. Coal, meanwhile, is projected to maintain its grip on more than half the world's power supply. Continued rises in coal and oil use would exacerbate urban air problems in industrialized cities that still exceed air pollution health standards and in megacities such as Delhi, Beijing, and Mexico City, which experience thousands of pollution-related deaths each year. And prolonging petroleum and coal reliance in transportation and electricity would increase annual global carbon emissions from 6.1 to 9.8 billion tons by 2020, accelerating climate change and the associated impacts of sea level rise, coastal flooding, and loss of small islands; extreme weather events; reduced agricultural productivity and water availability; and the loss of biodiversity.

Hydrogen cannot, on its own, entirely solve each of these complex problems, which are affected not only by fuel supply but also by

---

Hydrogen fuel cells could also help address global energy inequities.

---

such factors as population, over- and under-consumption, sprawl, congestion, and vehicle dependence. But hydrogen could provide a major hedge against these risks. By enabling the spread of appliances, more decentralized "micropower" plants, and vehicles based on efficient fuel cells, whose only byproduct is water, hydrogen would dramatically cut emissions of particulates, carbon monoxide, sulfur and nitrogen oxides, and other local air pollutants. By providing a secure and abundant domestic supply of fuel, hydrogen would significantly reduce oil import requirements, providing the energy independence and security that many nations crave.

Hydrogen would, in addition, facilitate the transition from limited nonrenewable stocks of fossil fuels to unlimited flows of renewable sources, playing an essential role in the "decarbonization" of the global energy system needed to avoid the most severe effects of climate change. According to the World Energy Assessment, released in 2000 by several United Nations agencies and the World Energy Council, which emphasizes "the strategic importance of hydrogen as an energy carrier," the accelerated replacement of oil and other fossil fuels with hydrogen could help achieve "deep reductions" in carbon emissions and avoid a doubling of preindustrial carbon dioxide concentrations in the atmosphere—a level at which scientists expect major, and potentially irreversible, ecological and economic disruptions. Hydrogen fuel cells could also help address global energy inequities—providing fuel and power and spurring employment and exports in the rural regions of the developing world, where nearly two billion people lack access to modern energy services.

Despite these potential benefits, and despite early movement toward a hydrogen economy, its full realization faces an array of technical and economic obstacles. Hydrogen has yet to be piped into the mainstream of the energy policies and strategies of governments and businesses, which tend to aim at preserving the hydrocarbon-based status quo—with the proposed U.S. energy policy, and its emphasis on expanding fossil fuel production, serving as the most recent example of this mindset. In the energy sector's equivalent of U.S. political campaign finance, market structures have long been tilted toward fossil fuel production. Subsidies to these energy sources—in the form of direct supports and the "external" costs of pollution—are estimated at roughly $300 billion annually.

The perverse signals in today's energy market, which lead to artificially low fossil fuel prices and encourage the production and use of those fuels, make it difficult for hydrogen and fuel cells—whose production, delivery, and storage costs are improving but look high under such circumstances—to compete with the entrenched gasoline-run internal combustion engines and coal-fired power plants. This skewed market could push the broad availability of fuel cell vehicles and power plants a decade or more into the future. Unless the antiquated rules of the energy economy—aimed at keeping hydrocarbon production cheap by shifting the cost to consumers and the environment—are reformed, hydrogen will be slow to make major inroads.

One of the most significant obstacles to realizing the full promise of hydrogen is the prevailing perception that a full fledged hydrogen infrastructure—the system for producing, storing, and delivering the gas—would immediately cost hundreds of billions of dollars to build, far more than a system based on liquid fuels such as gasoline or methanol. As a result, auto and energy companies are investing millions of dollars in the development of reformer and vehicle technologies that would derive and use hydrogen from these liquids, keeping the current petroleum-based infrastructure intact.

This incremental path—continuing to rely on the dirtier, less secure fossil fuels as a bridge to the new energy system represents a costly wrong turn, both financially and environmentally. Should manufacturers "lock in" to mass-producing inferior fuel cell vehicles just as a hydrogen infrastructure approaches viability, trillions of dollars worth of assets could be wasted. Furthermore, by perpetuating petroleum consumption and import dependence and the excess emission of air pollutants and greenhouse gases, this route would deprive society of numerous benefits. Some 99 percent of the hydrogen produced today comes from fossil fuels. Over the long run, this proportion needs to be shifted toward renewable sources, not maintained, for hydrogen production to be sustainable.

In the past several years, a number of scientists have openly challenged the conventional wisdom of the incremental path. Their research suggests that the direct use of hydrogen is, in fact, the quickest and least costly route—for the consumer and the environment—toward a hydrogen infrastructure. Their studies point to an alternative pathway that would initially use the existing infrastructure for natural gas—the cleanest fossil fuel and the fastest growing in terms of use—and employ fuel cells in niche applications to bring down their costs to competitive levels, spurring added hydrogen infrastructure investment. As the costs of producing hydrogen from renewable energy fell, meanwhile, hydrogen would evolve into the major source of storage for the limitless but intermittent flows of the sun, wind, tides, and Earth's heat. The end result would be a clean, natural hydrogen cycle, with renewable energy used to split water into oxygen and hydrogen, with the latter used in fuel cells to produce electricity and water—which then would be available to repeat the process.

There are no major technical obstacles to the alternative path to hydrogen. As one researcher has put it, "If we really decided that we wanted a clean hydrogen economy, we could have it by 2010." But the political and institutional barriers are formidable. Both government and industry have devoted far more resources to the gasoline- and methanol-based route than to the direct hydrogen path. Hydrogen receives a fraction of the research funding that is allocated to coal, oil, nuclear, and other mature, commercial energy sources. Within energy companies, the hydrocarbon side of the business argues that oil will be dominant for decades to come, even as other divisions prepare for its successors. And very little has been done to educate people about the properties and safety of hydrogen, even though public acceptance—or lack thereof—will in the end make or break the hydrogen future.

The societal and environmental advantages of the cleaner, more secure path to hydrogen point to an essential—and little recognized—role for government. Indeed, without aggressive energy and environmental policies, the hydrogen economy is likely to emerge along the more incremental path, and at a pace that is inadequate for dealing with the range of challenges posed by the incumbent energy system. Neither market forces nor government fiat will, in isolation, move us down the more direct, more difficult route. The challenge is for government to guide the transition, setting the rules of the game and working with industry and society toward the preferable hydrogen future.

This catalytic leadership role would be analogous to that played by government in launching another infrastructure in the early years of the Cold War. Recognizing the strategic importance of having its networks of information more decentralized and less vulnerable to attack, the U.S. government engaged in critical research, incentives, and public/private collaboration toward development of what we now call the Internet. An equally, and arguably even more, compelling case can be made for strategically laying the groundwork for a hydrogen energy infrastructure that best limits vulnerability to air pollution, energy insecurity, and climate change. Investments made today will heavily influence in what manner and how fast the hydrogen economy emerges in coming decades. As with creating the Internet, putting humans on the moon, and other great endeavors, it is the cost of inaction that should most occupy the minds of our leaders now, at the dawn of the hydrogen age.

IV. Energy and the
Global Environment

Renewable Energy Consumption in Developing Asia, 1999, 2010, and 2020

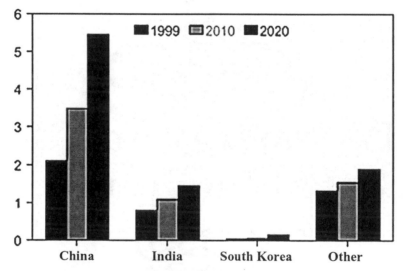

Sources: 1999: Energy Information Administration (EIA), *International Energy Annual 1999*, DOE/EIA-0219(99) (Washington, DC, January, 2001). 2010 and 2020: EIA, World Energy Projection System (2001).

Editor's Introduction

E nergy resources are vital to societies and nations. They are the engines of growth and barometers of power. It is not surprising, then, that many of the conflicts of the past century have been, at least on some level, over guaranteeing access to energy supplies, in particular oil.

The first few articles gathered in this section consider the impact of this most sought-after natural resource on its consumers and suppliers. America is the world's greatest oil consumer, and the stuff is considered so vital to national security that there is an emergency supply stored in huge underground salt caverns along the coastline of the Gulf of Mexico. Writing two months after the terrorist attacks on the World Trade Center and the Pentagon, David and Arlie Hochschild argue that dependency on oil imports has historically enmeshed the U.S. in the affairs of repressive regimes, and may bring further problems in Central Asia, the last big source of oil left on the planet. Americans can show their patriotism, they suggest, by weaning themselves from dependence on oil through conservation and alternative fuels. Just as Americans were asked during World War II to curtail their purchase of tin to aid the war effort, Americans today could be asked to drive fuel-efficient cars instead of gas-guzzling SUVs, for example.

Expanding on such ideas, Paul Epstein and Ross Gelbspan advocate a "global public works program" to create a clean energy infrastructure. An international program to create alternative sources, they suggest, would reduce the vulnerability of our traditional energy grid (nuclear power plants can be hit with planes; oil pipes can be blown up), redress economic inequalities among nations, and create new jobs and markets.

Writing at the start of the American-led war against the Taliban government in Afghanistan, Andy Rowell analyzes the strategic importance of Afghanistan and the rest of Central Asia's natural gas and oil reserves. Afghanistan, in particular, offers a potential oil transport route between the landlocked Caspian region and the Arabian sea, and it is thus in the interest of Western nations for there to be a friendly, stable government in power in Kabul.

In "The Real Price of Oil," James Surowiecki mounts a counterintuitive argument: countries in the Middle East and northwestern Africa remain poor precisely because they are rich in oil. Since the profit margin for drilling oil is so high, such countries have no incentive to invest in education for their people, develop technological resources, or diversify industries.

Douglas Farah's descriptions of the nightmarish scenes around the Niger River Delta in the next article provide concrete evidence of the costs of oil. In the heart of Nigeria's oil country, some 7 million residents live in abject poverty. Oil production has disrupted the Nigerians' traditional modes of hunting, fishing, and farming, and oil spills have despoiled the fragile wetland environment. After pressure from activists, the multinational oil companies and the newly democratic Nigerian government have begun returning some of their profits to the people of the region. Nigeria is expected to double or even triple oil output in the next decade, and it remains to be seen whether the Nigerian people can make oil work for them.

"Power to the Poor" reports on the millions worldwide who live on inadequate or polluting fuel sources—the energy poor. Two of the world's 6 billion don't have electricity, for example, and the health of millions is jeopardized by poor, choking air. In this article, *The Economist* writer suggests that, in addition to altruism, there is a "solid commercial motive" for helping bring clean, reliable energy to the poor: access to energy will help the poor begin to help themselves. In the last article, Daniel Kammen, a professor of energy and society at the University of California at Berkeley, assesses current programs aimed at bringing renewable and clean energy technologies to the poor. He looks in particular at programs that have brought wind energy systems to Inner Mongolia and high-efficiency cookstoves to Kenya. Such programs demonstrate the great potential for small-scale, targeted energy development.

Hooray for the Red, White, Blue and Green[1]

BY DAVID HOCHSCHILD AND ARLIE HOCHSCHILD
LOS ANGELES TIMES, NOVEMBER 11, 2001

In 1942, in the aftermath of Pearl Harbor, Americans began to understand that in order to win the war and to live again in a world at peace, they would have to make great personal sacrifices on the "home front." America began producing tanks, planes and reduced or stopped producing many household items such as toasters and waffle irons made with materials needed for the war.

The government was specific in what it asked citizens to give up, and it spelled out what their altruism would mean to the war effort. In his book *Don't You Know There's a War On?* Richard R. Lingeman describes the kinds of trade-offs put before the American public. "If every family in America would forgo buying one [tin] can a week, this would save 2,500 tons of tin and 190,000 tons of steel—the equivalent of 5,000 tanks or 38 Liberty ships. Thirty thousand razor blades contained enough steel to make fifty 30-caliber machine guns." Bernard Baruch, appointed by President Franklin D. Roosevelt to get Americans to conserve essential resources, rationed gas and lowered speed limits to 35 miles per hour on national highways.

People didn't just do without; they did things differently. Victory gardens—20,000,000 of them in parks, vacant plots and back yards—produced 40% of all the vegetables grown in the country. People also changed their tastes. Because rubber was scarce, companies stopped making women's rubber girdles, and fashion designers created new dress styles in response. In essence, during World War II, Americans saved, substituted, recycled and proudly did with less. They invented the idea of "green"—before they had the term—and put it together with red, white and blue.

The September 11 attacks have launched us into a very different kind of war. And it may be time now to take a leaf from our history and respond in the same spirit. Many Americans would like to help out—if they only knew how.

They also need to know why. The main reason is this: we're in a war started by Osama bin Laden but linked—profoundly and more lastingly—to oil. Since the 1970s, both our consumption of oil and our dependence on foreign oil have risen. Today, more than half of the oil consumed in the United States comes from overseas, much

---

1. Article by David Hochschild and Arlie Hochschild from *Los Angeles Times* November 11, 2001. Copyright © David Hochschild and Arlie Hochschild. Reprinted with permission.

of it from the Persian Gulf. We consume 25% of the oil produced in the world each year, yet have only 2% of the world's known reserves. This oil dependency enmeshes us with repressive regimes and countries locked in struggles between harsh oligarchies and fundamentalist rebels. Our involvement with such a regime in Iran ended disastrously, and a similar story may lie ahead in Saudi Arabia.

Indeed, our oil dependence may lead us to more unwanted conflicts in Central Asia. The next big source of oil is Central Asia, where the regimes are, if anything, harsher, and where our presence may create even more unrest. U.S. oil companies are already there, cutting deals and planning pipelines—aligning us with unsavory regimes in ways that make ordinary people hate us. In the end, our entanglements may make costly military interventions inevitable. In 1998, Dick Cheney, then CEO of Halliburton Co.—which during his watch landed $2.3 billion in U.S. government contracts or taxpayer guaranteed loans to drill oil—said, "I can't think of a time when we've had a region emerge as suddenly to become as strategically significant as the Caspian." The major way to get Caspian oil and gas out to where it can be loaded on ocean-going tankers is through Afghanistan.

Perhaps the most meaningful and lasting contribution Americans can make to the anti-terrorism efforts is to break the oil habit.

So, while many factors led up to September 11, oil is one of them. And if oil remains our primary energy source of choice, we can look forward to future hot spots in Uzbekistan, Tajikistan, Kazakhstan and Turkmenistan.

A shift toward renewable energy and conservation can also help reduce our vulnerability to terrorist attacks. Thank God, people are saying, that the hijacked planes didn't hit one of America's 109 nuclear power plants. But next time they could. Oil pipelines can be sabotaged and tankers sunk as well.

Given all this, perhaps the most meaningful and lasting contribution Americans can make to the anti-terrorism efforts is to break the oil habit. We cannot drill our way to an oil-based energy independence even if we try. All the oil in the Arctic National Wildlife Refuge would only yield enough to supply the nation for six months. Holland, Denmark and Norway have all set themselves the goal of achieving energy independence based on renewable energy, and we could do the same. We could declare it our "home front."

It will be a tough job. We will have to break some habits, but we won't have to change our values. Americans already believe in conservation, in green technology and in living within the globe's means. According to a recent ABC poll, 78% of Americans would like to see more energy conservation. Some 80% support more solar and wind power. We believe in conservation and renewable energy; we just don't yet act on our beliefs.

In a similar way 40 years ago, most Americans believed smoking led to cancer, emphysema and heart disease. Still, many of us kept smoking in restaurants, airplanes, workplaces and homes. In the last 40 years, Americans have cut their tobacco consumption in half, and lung cancer, heart disease and emphysema death rates have plummeted. It wasn't easy. It took education, legislation and litigation. We went up against a big industry. But we did it—and are doing it still.

In the same spirit, we can "green" our way to energy independence—by conserving energy and by generating the energy we need from renewable sources such as solar and wind power. The Bush administration will not lead on this; both President Bush and Vice President Cheney have made huge fortunes in oil. National Security Advisor Condoleezza Rice not only served on Chevron's board of directors; in 1993 the company actually named an oil tanker after her.

But one of the beauties of patriotic action on the home front is that we don't need to wait for leadership from on high: We can all lead. We can commit to personal acts of conservation like replacing gas-guzzling cars with fuel-efficient ones, taking public transportation or insulating our houses.

Legislators can travel by public transportation. Movie stars can dismount at premieres from hybrid cars. Enlightened corporate CEOs can erect wind turbines on company property—with flags on top. School boards can order solar panels for school roofs and link environmental curricula to them.

State and local governments can pass new laws. Last Tuesday, San Francisco voters approved a $100-million revenue bond—the largest municipal energy bond measure in the country—to purchase solar panels and wind turbines to be placed on city property. Ultimately, the expenditures will be entirely offset by energy savings. Since 1980, the cost of solar energy has decreased by 71% and the cost of wind energy by 89%. The more people move to solar and wind power, the greater the economies of scale, and the less it will cost.

In recent weeks, Americans have contributed hundreds of millions of dollars to the victims of September 11 and lined up for hours to donate blood. We're a good people. By using our most precious resources—our can-do spirits, our good will—we can achieve energy independence through clean and renewable energy. The seeds of today's victory gardens are already in our hands, but this time the "gardens" may be on our roofs.

A Worldwide Goal of Clean Energy[2]

BY PAUL R. EPSTEIN AND ROSS GELBSPAN
BOSTON GLOBE, OCTOBER 24, 2001

The United States has been traumatized since September 11 by the realization of just how vulnerable our society is. As shock yields to perspective, we face three simultaneous—and superficially unconnected—challenges. We must reestablish security, revitalize an eroding global economy, and address an increasingly unstable environmental climate. While their causes appear as disparate as their symptoms, they are all ultimately susceptible to a common solution, the first part of which involves a properly financed, global transition to clean energy sources. A worldwide transition to clean energy would reduce the significance of oil. Oil dependency has altered power relations among nations, skewed incomes within nations and financed terrorism. A renewable energy economy, with distributed home- and industry-based fuel cells, small hydro dams, windfarms and stand-alone solar systems, would make the nation's electricity grid a far less strategic target for future terror attacks. A properly-funded, U.S.-led global energy transition would also begin to redress the inequities that are splitting humanity into rich and poor. Runaway economic inequities are destabilizing the global political environment just as runaway carbon concentrations are destabilizing the global climate.

Many of our economic policies no longer work. The precipitous bursting of the consumer bubble, compounded by the new climate of fear, inhibits investments. We need a new set of policies to propel the global economy while preventing it from undermining the environment on which it depends. Serious recessions, or depressions, may need more than tax cuts or interest rate reductions. Most economists conclude that a protracted war against terrorism will not significantly lift the economy. It is time for governments to take the lead by instituting aggressive public works programs as part of economic stimulus packages. A global public works program to construct a clean energy infrastructure would create millions of jobs all over the world. It would raise living standards abroad without compromising ours. It would allow developing countries to grow without regard to atmospheric limits—and without the budgetary burden of imported oil. Former World Bank executive director Morris Miller has pointed out that energy investments in poor countries create more wealth per dollar than investments in any other sector. Just as the Marshall Plan revitalized European economies after World War II, this type of plan today would turn impoverished countries into robust trading partners.

It would represent a U.S. policy that is expansive, inclusive and cooperative. Around the planet, the deep oceans are warming, the tundra is thawing (jeopardizing Alaskan pipelines), glaciers are melting, infectious diseases are migrating and the timing of the seasons has changed. All that has resulted from one degree of warming. Earth is projected to warm from 4 to 10 degrees over this century, and weather patterns are expected to become more severe and uncertain if we do not act now. To stabilize the climate, humanity must reduce carbon emissions by about 70 percent, according to the Intergovernmental Panel on Climate Change. The researchers calculated that the world must derive half its energy from non-carbon sources by 2018 or the atmospheric concentrations of heat-trapping carbon will quadruple early next century. That would clearly be catastrophic. The answer to all these problems lies in a rapid switch away from oil and coal and to renewable energy. One set of three inter-active strategies include: Redirecting

---

The deep oceans are warming, the tundra is thawing . . . , glaciers are melting, infectious diseases are migrating and the timing of the seasons has changed.

---

the $200 billion industrial countries currently spend on subsidies for fossil fuels to renewable technologies to help the major oil companies transform themselves into renewable energy companies; Creating a fund of about $300 billion a year to transfer renewable energy to the developing world. Switching dollars from the financial sector into production could be done in a number of ways. One method is through a tax on international currency transactions, which today total $1.5 trillion per day. A tax of a quarter-penny per dollar on those transactions would yield $300 billion a year for that transfer. Alternatively, a carbon tax in industrial countries of about $50 per ton of carbon emissions would raise an equivalent amount. Requiring the parties to the Kyoto Protocol (including the United States) to increase their fossil fuel efficiency by 5 percent a year until the goal of 70 percent is met. British Prime Minister Tony Blair said that, even as we confront the networks of terror, we must also address our long-term common problems of global poverty and global climate change. A transition to clean energy would dramatically expand the total wealth in the global economy, pacify our inflamed climate and enhance our long-term security. It would extend the baseline conditions for peace among people and peace between people and nature.

Route to Riches: Afghanistan[3]

By Andy Rowell
Guardian, October 24, 2001

As the war in Afghanistan unfolds, there is frantic diplomatic activity to ensure that any post-Taliban government will be both democratic and pro-West. Hidden in this explosive geo-political equation is the sensitive issue of securing control and export of the region's vast oil and gas reserves.

The Soviets estimated Afghanistan's proven and probable natural gas reserves at 5 trillion cubic feet—enough for the U.K.'s requirement for two years—but this remains largely untapped because of the country's civil war and poor pipeline infrastructure.

More importantly, according to the U.S. government, "Afghanistan's significance from an energy standpoint stems from its geographical position as a potential transit route for oil and natural gas exports from central Asia to the Arabian Sea."

To the north of Afghanistan lies the Caspian and central Asian region, one of the world's last great frontiers for the oil industry due to its tremendous untapped reserves. The U.S. government believes that total oil reserves could be 270bn barrels. Total gas reserves could be 576 trillion cubic feet. These dwarf the U.K.'s proven reserves of 5bn barrels of oil and 27 trillion cubic feet of natural gas.

The reason oil is so attractive to the U.S.—which imports half of its oil—and the West, is for three reasons. "Firstly, it is non-OPEC oil," says James Marriott, an oil expert from Platform, an environmental NGO. "OPEC has been the bete-noire of the West since its inception in 1960. Secondly, these states are not within the Arab world and thirdly, although they are Muslim, they are heavily secularised."

The presence of these oil reserves and the possibility of their export raises new strategic concerns for the U.S. and other western industrial powers. "As oil companies build oil pipelines from the Caucasus and central Asia to supply Japan and the West, these strategic concerns gain military implications," argued an article in the *Military Review*, the journal of the U.S. army, earlier in the year.

Despite this, host governments and western oil companies have been rushing to get in on the act. Kazakhstan, it is believed, could earn $700bn (pounds 486bn) from offshore oil and gas fields over the next 40 years. Both American and British oil companies have struck black gold. In April 1993, Chevron concluded a $20bn joint venture

---

3. Article by Andy Rowell from the *Guardian* (*www.guardian.co.uk*) October 24, 2001. Copyright © Andy Rowell. Reprinted with permission.

to develop the Tengiz oil field, with 6–9bn barrels of estimated oil reserves in Kazakhstan alone. The following year, in what was described as "the deal of the century", AIOC, an international consortium of companies led by BP, signed an $8bn deal to exploit reserves estimated at 3–5bn barrels in Azerbaijan.

The oil industry has long been trying to find a way to bring the oil and gas to market. This frustration was evident in the submission by oil company Unocal's vice-president John Maresca, before the U.S. House of Representatives in 1998. "Central Asia is isolated. Their natural resources are landlocked, both geographically and politically. Each of the countries in the Caucasus and central Asia faces difficult political challenges. Some have unsettled wars or latent conflicts."

The industry has been looking at different routes. The Caspian Pipeline Consortium (CPC) route is 1,000 miles west from Tengiz in Kazakhstan to the Russian Black Sea port of Novorossiisk and

---

"Washington's attitude towards the Taliban has been, in large part, a function of oil."—Steve Kretzmann, Institute for Policy Studies

---

came on stream last week. Oil will go by tanker through the Bosporus to the Mediterranean. Another route being considered by AIOC goes from Baku through Tbilisi in Georgia to Ceyhan in Turkey. However, parts of the route are seen as politically unstable as it goes through the Kurdistan region of Turkey and its $3bn price tag is prohibitively expensive.

But even if these pipelines are built, they would not be enough to exploit the region's vast oil and gas reserves. Nor crucially would they have the capacity to move oil to where it is really needed, the growing markets of Asia. Other export pipelines must therefore be built. One option is to go east across China, but at 3,000km it is seen as too long. Another option is through Iran, but U.S. companies are banned due to American sanctions. The only other possible route is through Afghanistan to Pakistan. This is seen as being advantageous as it is close to the Asian markets.

Unocal, the U.S. company with a controversial history of investment in Burma, has been trying to secure the Afghan route. To be viable Unocal has made it clear that "construction of the pipeline cannot begin until a recognised government is in place in Kabul that has the confidence of governments, lenders, and our company."

This, it can be argued, is precisely what Washington is now trying to do. "Washington's attitude towards the Taliban has been, in large part, a function of oil," argues Steve Kretzmann, from the Institute for Policy Studies in the U.S. "Before 1997, Washington refused to criticise and isolate the Taliban because Kabul seemed to favour Unocal, to build a proposed natural gas pipeline from Turkmenistan through Afghanistan to the Pakistan coast."

In 1997, the Taliban signed an agreement that would allow a proposed 890-mile, $2bn natural gas pipeline project called Centgas led by Unocal to proceed. However, by December 1998, Unocal had pulled out citing turmoil in Afghanistan making the project too risky.

To secure stability for the Afghan pipeline route, the U.S. State Department and Pakistan's intelligence service funnelled arms to the Taliban, argues Ahmed Rashid in his book: *Taliban: Militant Islam, Oil and Fundamentalism in Central Asia*, the book Tony Blair has been reportedly reading since the conflict started. Rashid called the struggle for control of post-Soviet central Asia "the new Great Game."

Critics of the industry argue that so long as this game is dependent on fossil fuels the region will remain impoverished due to the effects of the oil industry, which is, says Kretzmann, "essentially a neo-colonial set-up that extracts wealth from a region. The industry is sowing the seeds of poverty and terrorism. True security, for all of us, can only be achieved by reducing our dependence on oil."

The Real Price of Oil[4]

By James Surowiecki
New Yorker, December 3, 2001

Call them—as the press has—"breeding grounds," "flashpoints," or "Powder kegs." Whatever name you use, the struggling economies of the Middle East and North Africa, where unemployment is routinely above fifteen per cent and economic growth is minimal, are now recognized to be potent sources of unrest, as a parade of Presidents and Prime Ministers argued two weeks ago at the U.N. General Assembly. Worrying about poverty, of course, means worrying about the causes of poverty, and in the past two months we've heard myriad explanations for Middle Eastern deprivation: Western exploitation, Arab corruption, Islamic hostility to modernization, globalization, the failure of globalization, and a dearth of Western aid. What hasn't been mentioned is oil.

Oil is usually thought of as a good thing—the only thing, in fact, keeping the Middle East and North Africa afloat. Without high oil prices, the argument goes, many of the countries in the region would sink into poverty and political chaos. But oil revenues are to the Middle East what heroin is to the junkie. Day to day, shooting up keeps you from feeling sick, over time, though, it keeps you from being healthy.

Countries like Saudi Arabia, Algeria, and Iran are textbook examples of what economists call "the resource curse." Though rich in natural resources, they are poor. Or, rather, they are poor in part because they are rich in natural resources. A 1995 study of ninety-seven developing countries by the economists Jeffrey Sachs and Andrew Warner found that the more important natural resources were to a country's economy, the lower its growth rate was. Of all the resource-rich countries they studied, only two were able to grow as fast as two per cent a year, while a host of the resource-poor nations grew much faster. "Just look around the world, and tick off the countries that are resource-rich," Warner says. "They are not rich countries, and they obviously haven't grown rapidly, because otherwise they would be." The biggest economic flops of the last decade—Russia, Argentina, Nigeria—abound in natural resources.

Why does the resource curse exist? The simplest answer is that being dependent on natural resources makes a country less likely to invest in other things that might be economically valuable, especially manufacturing. In part, this is a question of prices. When a country like Saudi Arabia is flush with natural-resource money, everything in that country becomes more expensive, including labor. It's difficult to open, say, a factory to make khakis for the Gap, because the khakis would be too expensive to compete with those made elsewhere. This hurts, because manufacturing, with its competitive pressures and its demand for technological innovation, is a key source of economic growth for developing countries. Natural resources are depleted over time, while the benefits of technological innovation actually increase.

> *Natural resources are depleted over time, while the benefits of technological innovation actually increase.*

Natural-resource wealth also depresses entrepreneurialism. It costs Saudi Arabia almost nothing to get a barrel of oil out of the ground, but that barrel, even now, can sell for seventeen dollars. It's a profit margin you'll find in almost no other business in the world. And it makes other business opportunities less attractive. People who might be inventing new products or opening shoe factories instead spend their days figuring out how to get a share of the oil money. "If you can get a reasonable income without really working, lots of people will take that," Howard Pack, an economist at the Wharton School, says. The problem is exacerbated by the fact that natural resources tend to be controlled by state-run monopolies, which pretty much insures a low level of innovation and competitiveness, and encourages people to look to the state, instead of themselves, for solutions. In Saudi Arabia, only twenty per cent of the jobs in the private sector are held by Saudis, who prefer government sinecures.

A dependence on natural resources fosters the illusion that you get rich by taking what's already there, rather than by creating something new. But the automobile, the electric turbine, and the computer chip were not there for the taking; they had to be created. There are countries that have recognized this and, in doing so, evaded the resource curse. Warner points to the example of Chile, which, despite vast copper fields, boomed in the nineteen-nineties. The growth rates of Malaysia and Indonesia over the past thirty years have far outpaced those of Middle Eastern states. And the little African nation of Mauritius became a powerhouse even though it started out with an economy that relied almost entirely on sugar exports.

These countries succeeded because they used their resource wealth to diversify their economies. They set up special export zones to encourage manufacturing. They countered the impact of high prices by devaluing their currencies. They opened their markets to free trade. And they invested heavily in education.

The oil-producing nations of the Middle East and North Africa, by contrast, have done none of these things. Their economies are still, for the most part, closed to the world. They have little or no manufacturing and, perhaps as a result, little or no technological innovation. They spend far more time fighting over how to divvy up the spoils than over how to create new wealth, which means that every economic decision becomes a political one. And they have invested very little in education. The oil producers are addicts. They prefer the comfortable squalor of staying hooked to the work it would take to kick the habit. Ultimately, the resource curse is less something they are afflicted with than something they have inflicted on themselves. Right now, the rest of the world is paying the price.

Nigeria's Oil Exploitation Leaves Delta Poisoned, Poor[5]

By Douglas Farah
Washington Post, March 18, 2001

The swamp and palm trees surrounding Well 19 are still black, seven months after thousands of barrels of crude oil spilled into the jungle and caught fire, fouling the water and scorching the tropical forest. Nearby, a stream of natural gas hisses from a pipe that has recently been sabotaged.

Not far away is an immense natural gas flare that shoots a flame 300 feet into the sky, noticeably raising the temperature at the nearby village of Oshie by several degrees.

The scenes are repeated all around the Niger River Delta, a fragile wetland of about 42,000 square miles that produces 2 million barrels of crude oil a day and that is worked by five multinational firms. It is home to about 7 million Nigerians. Abandoned by the government, hostile to the oil companies and ecologically ravaged, the delta is in a dismal state that sometimes seems impossible to remedy, one perpetuated by a seemingly endless cycle of distrust and violence.

By any measure, the delta is an environmental basket case.

Whether caused by carelessness, human error or sabotage, oil spills have dumped at least 2.5 million barrels of oil—equal to 10 *Exxon Valdez* disasters—into the delta from 1986 to 1996, according to a recent unclassified study commissioned by the CIA. Oil companies acknowledge that at least 100,000 barrels were spilled in 1997 and 1998.

And every day, 8 million cubic feet of natural gas are burned off in flares that light the skies across the delta, not only driving off game, hurting the fishing and poisoning the agriculture, but contributing to global warming.

The CIA study found that while oil extraction has "generated immense profits, the delta's inhabitants have suffered increasing poverty and a general decline in the quality of their lives due, in part, to the environmental impact of oil extraction. Corruption and bureaucratic incompetence have led to an almost total absence of schools, good drinking water, electricity or medical care."

Around Eriemu, ethnic Ijaw communities blame the oil companies for the August spill and months-long delay in cleaning up. The villagers, who said cleanup efforts did not begin until last month, want

---

economic compensation for the ruined lands and water, as well as more oil-company investment in health, education and water systems.

"We will not be able to use this land for the next 25 or 30 years at best," said Peter Akpagra, at Eriemu village. "There is no way all that oil can be cleaned up—Shell can't do it—so our farming has stopped."

Such disputes have often spilled over into violence, with young people in the communities taking oil workers hostage, occupying pumping stations or sabotaging pipelines, and oil companies relying on the often brutal and corrupt Nigerian military and police to maintain order, sometimes even paying military and police salaries in the region.

Ijaw leaders have taken strong public stands against such violence. But they said the jarring sight of foreign oil camps with running water, electricity and health care beside the villages with none of those amenities has led to more and more radical actions by disaffected youths.

"The oil companies must and should be subordinate to the people."—Oranto Douglas, deputy director of Environmental Rights Action

---

"The oil companies must and should be subordinate to the people," said Oranto Douglas, deputy director of Environmental Rights Action, a Nigerian advocacy group. "Right now they are lords and masters."

The Shell Petroleum Development Corp., Shell's Nigerian branch that runs the site and is by far the largest oil producer in the delta, says the problem at this wellhead is not so simple.

Executives said there can be no compensation because the well was sabotaged and the cap ripped off, something community leaders reluctantly acknowledged. Compensation, they said, would lead to more sabotage.

And, said Shell executives, the community blocked cleanup efforts in an effort to force the company to pay, leaving the oil to soak into the water and land.

Shell also argues that, while it funds some regional development, it is up to the government, not the company, to provide basic services.

Hubert Nwokolo, Shell's general manager of development, said that the company has radically altered its approach and raised its community development spending from about $10 million in 1996 to $55 million last year in 150 communities.

But Shell officials argue that, while they have an obligation to the communities, so does the government, which is largely absent. They said they were trying to get the government to use its own oil revenue to establish services in this region.

"If it did, the pressure would be much less on the oil companies," Nwokolo said. "We build schools, put in water systems and electricity. But we also pay our taxes. It is really the state's job to take care of its people."

Douglas and other activists say that while there is some truth to that argument, the oil companies have often functioned as virtual arms of the government. In doing so, they have developed close relations with the brutal and corrupt military regimes that ruled Nigeria until 22 months ago.

Shell has been forced to shut down its oil production in the eastern Ogoni region since 1995, for example, when ethnic Ogoni leaders and youths carried out a campaign of violence against the company

---

The democratic government of President
Olusegun Obasanjo . . . has slowly begun
fulfilling its constitutional obligation to give
13 percent of the country's oil revenue back to
the six Niger Delta states.

---

after the execution of Ogoni leader and environmentalist Ken Saro-Wiwa. The executions were carried out by the military government under dictator Sani Abacha, and the Ogonis alleged that Shell did not use its influence with the government to try to free Saro-Wiwa.

"The Niger Delta is not a law-and-order question," Douglas said. "It is primarily and almost purely political and a question of survival. The communities want to protect their air, water and forest."

There are some small signs that the situation could be improving. Violent incidents such as kidnapping and sabotage dropped sharply last year. Activists such as Douglas and Moffia Akobo, a former oil minister who now heads the Southern Minority Movement here, who were driven underground during the years of military rule, now operate openly.

The democratic government of President Olusegun Obasanjo, who took office in May 1999, has slowly begun fulfilling its constitutional obligation to give 13 percent of the country's oil revenue back to the six Niger Delta states to use for development. Under the military, 1 percent was allocated, but even that didn't make it.

The first oil money was disbursed to the states in January, and government officials, activists and oil company executives say it has given the local and state governments an incentive to keep the oil flowing.

In addition, the government has formed the Niger Delta Development Corp., which is supposed to use a new 3 percent tax on oil companies to fund regional development. Since its establishment four months ago in Port Harcourt, the largest city in the delta, the corporation still has no working telephones and very limited office space. But as Akobo and other community leaders are quick to point out, almost every Nigerian government has set up some version of a development corporation here and most of the money has simply disappeared into the pockets of officials.

Power to the Poor[6]

ECONOMIST, FEBRUARY 10, 2001

After decades of intensive and expensive efforts to help the "energy poor" in developing countries, there is little to show for it all. As the World Energy Assessment (WEA), a joint effort by the United Nations and the World Energy Council, recently pointed out, "The current energy system is not sufficiently reliable or affordable to support widespread economic growth. The productivity of one-third of the world's people is compromised by lack of access to commercial energy, and perhaps another third suffer economic hardship and insecurity due to unreliable energy supplies."

That is clearly a big problem for poor countries. Increasingly, though, energy poverty is a matter of concern for rich countries, too, and it is in their interest to help establish a sustainable energy future for all the world's inhabitants.

If nothing else, they might feel a moral obligation. Development experts estimate that some 2 billion people, chiefly in the rural areas of poor countries, lack access to electricity. But the IEA thinks that the real number may be considerably higher, because "access to electricity" is often defined as a grid extension to a village, when in many villages only a handful of people actually have access to that power. In urban areas, too, the high cost of connection prevents many households from gaining safe access to electricity.

This does not mean, of course, that they do not use energy, but that they use it in its least convenient forms—eg, charcoal, crop residues, cow dung—and usually in ways that are damaging to both human health and the environment. Such inferior fuels make up perhaps a quarter of the world's total energy consumption, and three-quarters of all energy used by households in developing countries.

According to a recent analysis by Richard Ackermann of the World Bank, the costs of using inferior fuels can be staggering: he found that the urban areas of China alone lose some 20% of potential economic output because of the effect on human health of dirty energy use. In India, indoor air pollution from dirty fuels causes as many as 2 million premature deaths a year, particularly among women and girls, who do most of the cooking.

But altruism apart, the rich world also has a solid commercial motive for caring about the poor. The developing countries' voracious appetite for energy will soon have a huge effect on the availability and cleanliness of the stuff the world over, and perhaps even on the stability of energy markets. Energy consumption in the rich world, both in absolute terms and on a per-head basis, has always dwarfed that in poor countries, but in the next few decades developing economies, especially India's and China's, will start to catch up. The IEA reckons that two-thirds of the increase in energy demand between 1997 and 2020 will come from poorer countries. If China and India rely heavily on antiquated technology to produce power from their plentiful local supplies of coal, they will surpass even the United States as the leading emitters of carbon dioxide within decades, negating any efforts by rich countries to curb global warming. Unless rich countries help poor ones leapfrog to greener technologies, the world could soon become a nastier place for everybody to live in.

Fortunately, there are several reasons to think that the future for the world's energy poor need not be as bleak as the past. One is the liberalisation of energy markets. Another, related one is the shift away from grandiose energy projects supported by international financial institutions and aid donors. The most powerful one, though, may be grass-roots activism in poor countries. Taken together, these forces suggest that the command-and-control, fossil-fuel-based power grid may be superseded in the future by a nimbler, more decentralised and cleaner energy infrastructure that is more likely to serve the needs of the poor.

Unless rich countries help poor ones leapfrog to greener technologies, the world could soon become a nastier place for everybody to live in.

The liberalisation of electricity markets now in progress in many developing countries is likely to make a big difference. Market-minded technocrats in many Latin American economies have encouraged competition in both the generation and the retail distribution of power. Argentina has also privatised YPF, its state-run oil giant, and even allowed its takeover by Repsol of Spain, its former colonial master. China has partially privatised its big oil companies. Indian states are now reforming power distribution in an effort to stop the large-scale pilferage and waste of power.

With a few exceptions (mainly China), all this may mean the death of the gigantic power project. In the past, central planners have lavished vast sums on building dams, and both nuclear and coal-fired power plants. Such projects were usually completed late and well over budget, had a much greater social impact than expected and turned out to be far less efficient than promised.

The introduction of market reforms in the production and delivery of power has injected a strong dose of reality. Most of the money for new power projects in developing countries now comes from the private sector, so projects must be financed on commercial terms—and energy investors increasingly favour small, efficient power plants fired by natural gas or other forms of distributed generation.

Liberalisation is also bringing down subsidies for fossil fuels, which will favour clean forms of distributed generation. This matters in rich countries, too, but the bias in favour of dirty energy sources is greater in many poor countries.

The View from Below

However, the best reason to think that a happier energy future awaits the world's poor comes from the grass roots. Contrary to conventional wisdom, people in poor countries do care about greenery and cleaner energy, and are prepared to pay for it. Local activists all over the developing world, encouraged by like-minded people in the rich world and linked by the Internet, are clamouring for less pollution.

> *Contrary to conventional wisdom, people in poor countries do care about greenery and cleaner energy, and are prepared to pay for it.*

To make this possible, those people will have to be helped to climb up what the WEA calls "the energy ladder": from simple biomass fuels to convenient, efficient fuels (usually liquid or gaseous) for cooking and heating, and to electricity for most other uses. But decades of experience show that governments alone, however generous or well-intentioned, cannot do the job without the market.

"The developing world is just littered with examples of energy projects that have failed because donors or governments did not think about how they will be maintained and paid for," explains Christine Eibs Singer of E&Co, a charity that finances renewable energy. The key to sustainability, she argues, lies in helping local entrepreneurs create markets for the energy services that the poor actually need and are willing to pay for, rather than what distant bureaucrats think is appropriate.

It is a widespread misconception that the poor cannot or will not pay for energy. The World Bank thinks that of the 2 billion people currently without access to modern energy, perhaps half are able to pay commercial rates for electricity; the remaining billion, reckons the agency, will need some government subsidy. Other estimates suggest that only 5–10% of those 2 billion people can afford to pay commercial rates for power, and that another 15% could probably afford to do so if credit were provided. But even these more conservative sums add up to a market of 300 million–400 million households.

Market-driven policies, topped up with targeted subsidies, should be able to reach many of the 1.3 billion people living in what economists call absolute poverty, with incomes of less than a dollar a day. According to development experts, households are usually willing to spend about a tenth of their monthly income on energy, including cooking fuel. Even for those in absolute poverty, who typically have a household income of $40–60 a month, that would amount to $4–6 a month. Ms. Eibs Singer insists that "clean energy services could be provided for this amount on a market-driven basis, especially if you can target government programmes and subsidies at this level."

There is ample evidence that the poor do pay, often heavily, for inefficient, dirty energy—say, from kerosene, candle wax and batteries. Indeed, they often pay more per kilowatt than do middle-class, urban households or wealthy farmers who benefit from heavily subsidised grid electricity. For example, families in Peru's remote highlands on average spend about $4 a month on candles. For a bit more, they could afford the much higher-quality power offered by village power units: experts say that local entrepreneurs can turn a profit by leasing out a small, 35-watt solar unit, enough to power two bulbs and a radio, for about $80 a year.

In Yemen, dozens of tiny private generators have sprung up to service households not reached by the inadequate grid system. Although these small operators charge quite steep prices, even poor households scrape together the money rather than live in the dark. Electricity penetration in Yemen now tops 50% of households, far higher than in most countries at comparable income levels.

E&Co tries to help locals to help themselves. It invests in start-up enterprises in developing countries that want to deliver clean micropower to villages; it also advises them on how to draw up business plans, sales and distribution strategies and so on. The Tata Energy Research Institute (TERI), an Indian think-tank, uses a similar approach. In villages across India, the group has helped to train local entrepreneurs in setting up marketing chains, arrange decentralised credit and finance, and link the villages with manufacturers of village power units.

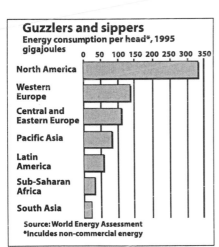

Guzzlers and sippers
Energy consumption per head*, 1995
gigajoules

Source: World Energy Assessment
*Inculdes non-commercial energy

To make such schemes work, the user must pay for the services, with targeted subsidies for the very poorest. Experience shows that giving the stuff away causes a massive waste of resources. Economic incentives are also needed for the private sector to maintain and improve those energy services. Local firms have tried various methods, ranging from cash up-front and pre-paid tokens to fee-for-service or leasing. The poor, it turns out, are usually excellent credit risks, though conventional banks in developing countries still refuse to accept this. Innovative microfinance initiatives, such as lending circles, generally see repayment rates of 92% to 98%.

Village Power, Meet Market Power

But although markets can indeed help the world's poorest, they offer no silver bullets. Such successful models as E&Co and TERI are not easy to scale up to the levels of funding that the big donors are used to. All the same, some of those donors are increasingly backing the private sector in its efforts to adopt a market-based approach. The United Nations Foundation's AREED (African Rural Energy Enterprise Development) project is helping on two levels: by supporting local enterprises with business development services and seed capital, and by training its local partners in developing countries to apply market principles. Another example is the Renewable Energy and Energy Efficiency Fund, a commercial equity fund sponsored by the International Finance Corporation (the World Bank's private-sector arm), which provides the investment-stage capital for renewable energy projects in developing countries. With the help of big western utilities, insurance companies and banks, the fund has raised some $65 million so far to boost renewable energy, with a focus on innovative village power schemes.

Grace Yeneza of Preferred Energy, a Philippine NGO, explains her group's work in several remote highland villages that had no access to grid electricity. Working with the *barangays*, or local councils, her group built a micro-hydroelectric plant on the creek separating the villages in order to deliver electricity to their common areas. Donor agencies paid for the equipment, but the villagers pitched in "equity" in the form of labour and local materials. They also organised themselves into a management committee to run the plant. Those who want power for their households must pay for it. The project has not only brought power to neglected villages, but co-operation among the villagers on the power project has also ended long-running squabbles over the local creek. The future for the world's poor may not be so dark after all.

Bringing Power to the People

Promoting Appropriate Energy Technologies in the Developing World[7]

By Daniel M. Kammen
ENVIRONMENT, JUNE 1999

More than two billion people worldwide (approximately 35 percent of the world's population) depend on traditional biomass fuels for the bulk of their energy needs. Many more rely on kerosene lanterns, diesel generators, and coal-fired power plants. However, the economic and environmental cost of these technologies is high. Small-scale (and often renewable) energy systems offer a viable alternative to these carbon-intensive energy sources that could both further sustainable development and improve human health. The research and development base, institutions, market policies, and training that will be needed to sustainably implement such systems are only beginning to appear, however.

This article examines the institutional capacity—for training, research, outreach, commercialization, and implementation—of a variety of organizations that are attempting to promote renewable and "appropriate" fossil fuel energy systems. The overall message is clear: While small-scale, decentralized systems can play a significant role in meeting the combined challenges of development and environmental conservation, there has been a general pattern of neglect of and under investment in such systems.[1] Although a number of organizations are focusing on these "mundane" technologies, they frequently suffer from inadequate financing and political support.

There is now an important opportunity for even very modest investments to produce disproportionately large returns. Sustained support for environmentally benign energy systems by U.S. energy, development, health, and environmental organizations and their foreign partners could do more for sustainable development and individual well-being than almost any other effort at international energy cooperation. At the same time, ill-conceived or half-hearted actions could do just as much to harm the further development of clean energy systems.

7. *Environment*, vol. 41, no. 5, pp. 11–15, 34–41, June 1999. Reprinted with permission of the Helen Dwight Reid Educational Foundation. Published by Heldref Publications, 1319 Eighteenth St., NW, Washington, DC 20036-1802. Copyright © 1999.

Potential Gains

Recent efforts to expand the use of traditional, small-scale fossil fuel and renewable energy technologies have resulted in dramatic improvements in performance, market power, sales and leasing opportunities, and end-user satisfaction in both developed and developing nations. Examples of these improvements include the growth of electricity minigrids using diesel or renewable energy sources, as well as the widespread introduction of improved cookstoves, photovoltaic solar home systems, wind turbines for household and small enterprise use, small hydroelectric generators, and biomass energy systems. Some of these technologies have already had a significant impact on local patterns of energy use, economic activity, and the environment.

In the Inner Mongolian Autonomous Region of China, for example, an estimated 130,000 small-scale (200- to 1,000-watt) wind energy systems were in operation in 1995, providing electricity to more than 500,000 people (about one-third of the population). The alter-

---

In the Inner Mongolian Autonomous Region of China, for example, an estimated 130,000 small-scale . . . wind energy systems were in operation in 1995.

---

native sources of power in this country are grid electricity (more than 90 percent of which is generated by coal) and stand-alone diesel generators. Thus, wind-generated electricity represents a substantial avoidance of greenhouse gas emissions while providing a basic service to a rural population.[2] The success of the program was achieved through careful planning and the creation of an effective regional and local infrastructure for manufacturing, sales, maintenance, and training. This included the development of a market for individual household systems through various subsidy mechanisms. The government of Inner Mongolia also recognized and allowed for the long lead time and sustained level of support necessary to create a thriving local industry.

The project has also led to technology transfers at many levels, both between the Inner Mongolian Autonomous Region and other countries and among various national, regional, and local organizations within China. Discussions are also under way in other parts of China and other Asian and Southeast Asian nations about replicating the program, but so far there is not enough institutional capacity to support such ventures despite strong interest by governments, nongovernmental organizations (NGOs), and the private sector.

Another prominent example is a more efficient ceramic cookstove (the *jiko*) developed in Kenya.[3] At least 700,000 such stoves are now in use in that country; more than 50 percent of urban homes possess one, along with roughly 16 percent of rural homes.[4] Some 200 small-scale businesses and artisans now produce in excess of 13,000 stoves each month. Both the stove itself and the general program for disseminating it have been adapted for use in a number of other African nations.

The process of research, development, demonstration, and commercialization that led first to the improved *jiko* and then to other high-efficiency stoves was seeded by international and local development funds. Importantly, policymakers decided not to directly subsidize the production and dissemination of these stoves but to provide support to designers and manufacturers. Because the stoves were relatively expensive (about $15) and their quality was highly variable, sales were slow at first. However, continued research and increased competition among manufacturers and vendors spurred innovations in both the materials used and the methods of production. There is now an extensive marketing network for these stoves, and prices have fallen to $1–3, depending on size, design, and quality. This is consistent with the "learning

---

Kenya . . . has the highest penetration rate of photovoltaic (PV) systems in the world.

---

curve" theory whereby the price of a new technology decreases by a uniform amount (often about 20 percent) for each doubling of cumulative sales.[5]

Kenya also has the highest penetration rate of photovoltaic (PV) systems in the world, with more than 80,000 systems in place and annual sales of approximately 20,000 systems.[6] Some 50 local and 15 international importers, assemblers, installers, and after-sales providers serve this market, which developed without significant aid, subsidies, or other forms of support. (For more details, see the sidebar at the end of this article.) While the current market is strong, there is still a tremendous need for the standardization of equipment as well as research on batteries, lighting fixtures, and electronic ballasts. In addition, possible credit arrangements need to be studied, as do the relative advantages of leasing a system rather than purchasing it. All these steps would immeasurably strengthen the local PV industry and further sustainable energy research generally.

Both the Inner Mongolian and Kenyan examples illustrate the great potential of small-scale, decentralized energy systems—as well as the need for institutional resources to handle the design, testing, and after-market (i.e., sales and service) issues surrounding these technologies. From the policy standpoint, two issues in particular need to be addressed: the capacity of current institutions, in collaboration with new partners, to implement important projects; and the opportunities for and constraints on donor or market-based programs to promote sustainable markets for alternative energy systems.

Current Institutional Capacity

A number of problems have plagued the institutions that support research on the implementation of small-scale and decentralized energy technologies: lack of steady funding; a paucity of training venues, technology and information exchange, and technology standards;[7] lack of financing for local commercialization efforts; and the reluctance of some governments to subject the national utilities to competition from stand-alone and or minigrid systems. However, two underlying institutional weaknesses have been poor communication and slow recognition that basic research is needed on these energy systems.

Communication Problems

Up until now, efforts to disseminate small-scale and decentralized energy systems have been largely isolated from one another. Countries have generally not been able to draw on other countries' experiences in this area or even on the lessons that other programs (such as agricultural extension) might have to offer. There are signs that this is changing, however, as a growing number of governmental and nongovernmental groups are now evaluating the full range of efficient, decentralized energy options. Such programs are critical to building a base for international collaboration. At the same time, some of these groups are not fully aware of the extensive technological developments, field applications, and innovative institutional arrangements that have taken place elsewhere in the world. The small and sporadic funding with which many researchers and practitioners have had to contend has led to overly specific project- and funding-driven investigations that do not encourage wide discussion and comparative analysis.

Development professionals, NGOs, academics, government policymakers, local utilities, and commercial energy providers rarely exchange project information, compare their efforts, or develop collaborative relationships. A dramatic example of the communications problem involves China. That country has the most widespread dissemination program for renewable energy technologies in the world. Yet its experience with more than 25,000 village-scale (less than 100 kilowatt) hydroelectric power stations, 150,000 household wind turbines, 120 million improved cookstoves, and tens of millions of small biomass, coal, and waste-powered boilers is largely unknown (and certainly understudied) in other countries.[8] Much of China's success in this area has been achieved through an enlightened mix of national standards and support for R&D coupled with efforts to encourage local innovation, heterogeneity, and entrepreneurial competition. Those who wish to learn about these efforts, however, can often only do so through interactions with China's somewhat balkanized bureaucracy, access to which frequently requires personal contacts.

The barriers to learning about China's programs and institutional capacity only heighten the impression, already widespread within the renewable energy community, that the Chinese experience is so unique that it cannot provide useful lessons for other countries. Ironically, China provides a wealth of relevant information on such important topics as effective low-subsidy dissemination approaches, institutional arrangements to promote market formation, local technology development strategies, and many others.[9] But little of this is published, much of the documentation is held by regional offices, and (as in many initiatives related to household energy) a lot of the project planning is done informally.

Scholarly attention to the problems of small-scale and decentralized energy systems is notable primarily for its absence.

While China may be an extreme example, there have been similar failures to communicate information about renewable energy experiences in Latin America, Africa, South Asia, and elsewhere. Even with the advances in information exchange stemming from the Internet, practitioners often assume that the barriers they face—and the solutions they come up with—are unique to their own setting. As surprising as it may seem, U.S. practitioners know relatively little about renewable energy activities in Canada and the United Kingdom.

Signs of change are beginning to appear, however. Recently, there have been a number of international conferences on topics such as small-scale energy technology and finance (for example, the Village Power conference series sponsored by the U.S. National Renewable Energy Laboratory and the World Bank has become a large and important meeting[10]). While participation in these meetings now includes an impressive range of governmental, NGO, multinational, business, academic, and advocacy groups, attendees come predominately from the United States and Europe. What is needed are comparative technical, economic, political, and commercial studies of the technologies and practices of different nations and regions. These efforts should be cooperative, and they should focus not only on particular projects or programs but also on the social and political contexts that shape them.

Absence of Scholarly Interest

Scholarly attention to the problems of small-scale and decentralized energy systems is notable primarily for its absence. There are very few research and teaching units in higher education devoted primarily to these topics. Much of the academic work that has been done has focused on the technology itself, not the social science aspects of adopting, adapting, and managing that technology. Only recently have the renewable energy units of national laboratories and academies of science in many countries begun to examine the commercialization of small-scale and decentralized technologies. As a consequence, issues such as the application of such technologies to rural development and ways to finance their dissemination have been left to institutions and individuals lacking the resources, the training, and in some cases, the inclination to carry out systematic studies. The absence of an institutional base for renewable energy studies is reflected in the fact that there are few prominent academic journals devoted to renewable energy per se and very limited venues in other academic publications. Furthermore, some of the articles that are published in the field lack rigor and detailed quantitative analysis.

The dissemination of renewable energy suffers from the lack of research in several ways. There is surprisingly little information on such important matters as long-term performance, the economic costs and benefits, the effectiveness of subsidies, and the social consequences of renewable energy. Interestingly, something very similar occurred in the case of the Green Revolution. For years, analysts paid little attention to the long-term results of planting the new, high-yielding varieties of rice and wheat; when they finally did so, they were very surprised to learn that average yields had actually declined in one of the few long-term experiments.[11]

In areas such as these, basic questions often go unanswered because the information needed is either not collected or relates to the largely undocumented informal and household energy sector. Neither businesses (which focus on profits) nor NGOs (which can be tied to short-term and intermittent funding cycles) have much incentive to conduct or support such research. Even governments may shy away from it out of a reluctance to critically review their own programs. Academia is thus the only real source of such information. As it happens, there are important precedents for academic research of this nature. Forestry departments, for instance, have long provided crucial information on the technical and social issues on which national timber policies are often based. Similar efforts are needed on behalf of small-scale and decentralized energy planning.

An area that particularly suffers from the lack of critical analysis is the relationship between renewable energy projects and the social and economic contexts in which they are embedded. All too often, projects are planned, implemented, and evaluated on the basis of unexamined assumptions about local conditions and the social and economic consequences of these projects. Broader discussions of the role of traditional renewable or stand-alone fossil fuel energy sources in rural development strategies are rare, despite their obvious importance to rural electrification and the impacts of linking marginalized areas to the formal economy.

The paucity of academic research on renewable energy sources for rural development stems from two factors, the lack of funding for such research and its lack of academic respectability. Because this subject does not fall within the purview of existing academic departments, it lacks access to many mainstream sources of research funds. At the same time, it does not really correspond to any of the funding categories now employed by governments and private donors. To some extent, this reflects the fact that the development of rural energy infrastructure is no longer a priority in rich countries. Nonetheless, it is surprising—with all the concern about global carbon emissions and the hopes that have been placed on renewable energy—that more funding for academic research in this area has not been forthcoming. Part of the challenge, of course, is for scholar-practitioners to demonstrate the need for and value of this work.

As a field of study, renewable energy for rural development suffers from the classic academic biases against interdisciplinary research. That is, it is too specific, too applied, and too far outside the concerns of any of the established disciplines. Sociologists find it too technical, engineers find it too "soft," and economists find it too marginal.[12] Energy studies in general are considered too applied for mainstream departments in many institutions, which tend to emphasize theory (as defined by disciplinary boundaries). For this reason, they are largely taken up only in applied and interdisciplinary programs such as environmental studies. Yet even within such programs, there is a strong bias towards theory (or fear of seeming insufficiently theoretical) that works against the publication of long-term case studies. Ironically, this bias against "mundane science" is a major impediment to the theoretical understanding of renewable energy, which is an inherently data-driven subject.

Some interdisciplinary fields, such as forestry, agricultural economics, and geography, are managing to overcome these same barriers. These fields have established methods that bridge the technical and social aspects of the problems that they study, along with publications that effectively report the findings of the field to a larger community. Small-scale, decentralized, and renewable energy studies as a field is at an early stage of this evolution. While it is not clear that a specific discipline is needed here, it is important to recognize the integrative and eclectic nature of these studies. A productive beginning would be the emergence of programs in "energy engineering" as well as in "business and energy," "energy, environment, and development" and "the political economy of energy."

At the same time, we should recognize the emerging research and policy projects that are working to fill this void. For example, the Renewable Energy Policy Project (REPP)[13] based in Washington, D.C. has conducted a far-reaching study of the potential for expanding the international market for photovoltaics. It concludes that what is needed is an integrated program to promote market analysis, involve government procurement, initiate a public information program, provide training opportunities, and "mainstream" photovoltaics in broader development policies. Similarly, the Renewables for Sustainable Village Power (RSVP)[14] program based at the National Renewable Energy Laboratory in Golden, Colorado, provides information on the implementation of renewable energy solutions for rural electrification programs in developing countries. RSVP's web site has several valuable components, including a project database, a library database, a discussion forum, an introduction to analytical models, a list of renewable energy contacts, information on issues in village power, and links to Internet

---

Crucial to the introduction of clean technologies is the development and testing of ways to overcome the initial cost barrier.

---

resources that offer a wide array of information. Both the REPP and RSVP initiatives are excellent examples of what is possible given a sustained and interdisciplinary approach to building institutional capacity for and analytic experience with small-scale and decentralized energy systems. The challenge is to support and sustain such capacity within developing nations, not as a copy of institutions in the North but as a uniquely southern institution.

Lessons from the South

A number of developing nations have programs devoted to studying traditional or alternative sources of energy. . . . Although our current capacity for investigating all of the issues surrounding such energy options is extremely limited—and is complicated by considerable differences in environmental, economic, political, and social conditions—some remarkably consistent lessons have emerged from the activities of these groups. One of the most fundamental is the need to begin with technologies that have been well researched and tested under local conditions. This is a basic but frequently overlooked step in the whole process of research, development, demonstration, and deployment.

Also crucial to the introduction of clean technologies is the development and testing of ways to overcome the initial cost barrier, which for new technologies is invariably highest for the first units installed. In the 1970s and 1980s, the adoption of many such technologies was financed by multinational donors, partly because the costs were often astronomical by local standards and partly because there was an emphasis on helping the neediest first. This tended to place the focus on poor rural populations, which had the least ability to pay for new technologies. As a result, many promising technologies were abandoned as premature or inappropriate after a short evaluation in the most demanding economic environment.

In the 1990s, the emphasis has largely shifted to market-based mechanisms, with subsidies usually limited to collateral functions such as infrastructure, training, and advertising and after-sale support. Direct subsidies are now often seen as the means of last resort, which tends to do more harm than good.[15] This approach has been productive because it focuses on the true costs and performance of new technologies. The *jiko* cookstove in Kenya offers a good example. Initial efforts to popularize it met with only limited success due to design problems and high cost.[16] The decision not to subsidize the purchase of these stoves but to support local designers, manufacturers, and vendors instead is now credited with laying the groundwork for a thriving and competitive industry. The introduction of improved cookstoves in China was very similar, with support given primarily to local research, entrepreneurial startup, and education/promotion programs.[17]

All the same, it is important to recognize that subsidies and grants can be important in initiating the research-to-commercialization chain or in promoting public awareness of and interest in new technologies. For example, the "Green Lights" program to promote the use of energy-efficient compact fluorescent lights (CFLs), which began in the United States and then spread to Poland, China, and other countries, was supported in each case by a subsidy.[18] However, these subsidies were targeted at CFL manufacturers, on the theory that the design and marketing improvements would be most efficiently passed on to consumers. Poland's Green Lights program is similar and has also been very successful.

The last general lesson for small-scale and renewable energy programs is the need to focus on market analysis, education, and development. Many of the groups involved in this area have cited the importance of involving and supporting local institutions; rationalizing taxes and tariffs; providing quality control and standards; and offering leasing, warranties, training, and other services to assist consumers.[19]

Sustaining the Markets

One area worth exploring in greater detail is the potential for expanding the markets for clean technologies. Efforts to develop markets for new small-scale and decentralized energy technologies are especially hampered by the relatively slow capital turnover rates in developing nations. . . . Lack of funds and the tendency to be cautious about new and untested technologies (particularly when they have to be imported) serve to dampen the market for new innovations. Broadening the market is even more difficult when these innovations will eliminate tasks such as collecting wood and making charcoal that are undervalued because they are done by women, children, or the elderly. Given barriers such as these, test marketing, promoting, and (if necessary) adapting the new technologies are particularly important. Such steps were an integral part of the process of introducing the *jiko* cookstove in Kenya, improved windmills in China, and efficient lights in Poland. Effective marketing is crucial in situations where even ideal innovations will fail on their own. In such situations, transforming the market should be coupled with the basic R&D process.

> *Energy prices generally do not reflect all the social costs imposed by pollution.*

Market Transformation Programs

Many policy analysts argue that governments should only support the development of new technologies, not their commercialization. There is a need, however, to broaden the definition of R&D to include dissemination and sustained use of new energy systems. This new view of R&D stems from the recognition that energy markets are not perfect: First, energy prices generally do not reflect all the social costs imposed by pollution; and second, private firms are generally unable to capture all the benefits from their investments in R&D. For both of these reasons, the private rate of return to R&D is less than the social rate, and left to their own devices, private firms will invest too little in this activity. The argument against support for commercialization is based on governments' past failures to "pick winners."[20] The failure of efforts such as the U.S. ethanol and synfuels programs, both of which invested heavily in specific technologies without understanding their market potential and the likely technological evolution in this area, only bolster this view. Nonetheless, there are compelling arguments for public funding of market transformation programs that subsidize energy innovations for the purpose of commercializing them.

One such reason is inherent in the production process itself. When a new technology is introduced, it is invariably more expensive than established alternatives. However, there is a clear tendency for the unit cost of manufactured goods to fall as the market expands and the producer gains experience in making them. The cost reductions are typically very rapid at first but taper off as the industry matures. When it accounts for all production costs this relationship is called an *experience curve*. It can be described by a progress ratio

(PR) where unit costs fall by *100*(1-PR)* percent with every doubling of production. (PR itself is defined as 1.0 less the rate of market expansion expressed as a decimal.) Typical PR values range from 0.7 to 0.9 and are applicable to any good that can be manufactured in quantity, such as toasters, microwave ovens, solar panels, and windmills.[21] For the PV industry, the value of PR is about 0.8, implying a 20 percent decrease in prices for every doubling of cumulative sales.

The benefits of production experience may accrue primarily to the first firm to enter the market, or they may spill over to its competitors. Spillovers could result, for instance, from hiring competitors' employees or reverse engineering their products (i.e., taking them apart to discover how they work), from informal contacts among employees of different firms, or even from industrial espionage. Such spillovers (both actual and potential) complicate markets considerably, generally leading to less production than would benefit society.

If a firm expects to retain the benefits of its production experience, it has two options. Because it will enjoy a growing cost advantage over potential competitors, it can restrict output—and thereby raise the price of the product—without inducing other producers to enter the market. In this case, the quantity produced will clearly be less than the social optimum. Alternatively, the firm can "forward price," producing at a loss initially to bring costs down and thereby maximize profits over the entire production period.[22] In principle, this would lead to the socially optimal level of production. In practice, however, it is unlikely that any firm would completely achieve that optimum. When spillovers are substantial, however, producers will generally not supply the socially optimal quantity of the good. Because they do not value the portion of their experience that benefits other firms, they do not forward price as much as they would otherwise.[23]

Regardless of the level of spillover, strong experience effects imply that output will be lower than socially desirable. That is, some consumers will be willing to pay more than the cost of additional production, but the market will not accommodate them. In this situation, market transformation programs can improve social welfare by eliminating the shortfall associated with experience effects.

When evaluating such programs, it is essential to account for the positive feedbacks between demand and production experience. By increasing the quantity produced in one year, market transformation programs reduce unit costs the next year and thus increase the quantity demanded in that year. This "indirect demand effect," in turn, augments production experience and lowers unit costs still further in future years. The process continues indefinitely, though it gradually tapers off after the market transformation programs are discontinued. Research has shown that indirect demand effects substantially raise the benefit-cost ratio of a typical market transformation program. Even without accounting for the environmental benefits, however, programs targeting photovoltaics and efficient lighting were found to have benefit-cost ratios above 1.0.[24] Quantifying the relative benefits of renewable energy and environmental programs in this way is important because in many circles they are seen as unlikely to be cost-effective.

In addition to correcting market imperfections, market transformation programs can reduce the social costs of pollution by promoting the use of clean energy technologies. The best antipollution policy, of course, is to fully internalize pollution costs through taxes or tradable emissions permits. Where this is politically impossible, however—and few governments have complete freedom in this area—market transformation programs can be an important part of a second-best solution.

A few caveats are in order here—the costs of poor program design, inefficient implementation, or simply choosing the wrong technology can easily outweigh the benefits of cost reductions. This suggests that market transformation programs should be limited to clean technologies with steep experience curves, high probabilities of market penetration once the subsidies are removed, and a price elasticity of demand of 1.0 or more.[25] Limiting market transformation programs to clean technologies enhances their performance by providing environmental benefits. The other two conditions ensure a strong indirect demand effect. Finally, as with energy R&D policy, public agencies should invest in a portfolio of new technologies to reduce the overall performance risk through diversification.[26]

Recommendations

This article has reviewed only some of the issues surrounding energy-related initiatives in the South (both successful and unsuccessful), but already seven specific recommendations have emerged:

Provide funding and other kinds of support to institutions rather than specific projects. Project-based funding has two drawbacks. First, it promotes technologies selected by donors, which usually fail to build institutional capacity or further local sustainability. Second, it often leads to boombust cycles for already small and underfunded organizations, further weakening their capacity to implement innovative solutions to local energy needs. As it happens, a number of developing nations have neglected (or been unable to pursue) locally appropriate solutions because international support was available only for a technology per se and not for human and institutional development.

Focus on capacity building and research, not just implementation. International funding agencies have usually neglected the first two of these goals,[27] even though they are arguably the most critical part of the R&D process for small-scale and decentralized technologies. Indeed, the adoption of such technologies requires much more attention than does the introduction of most large-scale energy systems.

Make institutional capacity building interdisciplinary, including economic, social, and policy research as well as the more traditional engineering and environmental studies. The small number of research groups and institutes focusing on small-scale and decentralized energy issues is a fundamental obstacle to building the necessary institutional capacity. Because these groups are small, however, a modest level of sustained international support with genuine local program direction could yield significant returns in capacity building.

Increase international support for regional networks. Networks such as the African Energy Policy Research Network, the Intermediate Technology Development Group, the Biomass Users Network, and others that build institutional capacity and cooperation (both South-South and South-North) have proven to be effective in furthering sustainable energy solutions.

Establish an interdisciplinary yet rigorous journal devoted primarily to small-scale and decentralized energy systems. Although several journals currently publish important material in this area (notably *Energy Policy, Energy for Sustainable Development, Environment*, and *Solar Energy*), these publications have yet to fully integrate policy analysis with empirical and theoretical studies that bring together the academic, practitioner, donor, and recipient perspectives. For that purpose, a journal like *Environmental Science and Technology*, which combines high-quality technical articles with policy pieces, reports on projects, current news, and professional advertisements, is needed. High-level support for academic as well as practitioner research and project documentation is also essential. Finally, it is critical that project managers have the opportunity to write up the results (positive or negative) of their efforts for presentation to the wider energy and development community.

Allocate subsidies appropriately. Subsidies for R&D and market development coupled with market mechanisms to set prices have been highly effective in fostering small-scale and decentralized energy industries.

Promote collaboration between the public and private sectors. The connections between governments, NGOs, and private firms are often very tenuous in developing nations. There are, however, a number of mechanisms that could be used to bridge this gap in the energy sector (e.g., policies to encourage technology leasing as well as research on energy services and preferences), which should be examined and supported by U.S. agencies and other organizations and their foreign partners.

Recent interest in environmental sustainability and the recognition that many renewable energy systems are nearing commercial viability has opened the door for significant international effort.

Research, development, and planning for small-scale and decentralized energy systems are an area where sound policy choices and resource allocation can have truly dramatic impacts.

Notes

1. See D. M. Kammen and M. R. Dove, "The Virtues of Mundane Science," *Environment*, July/August 1997, 10.
2. J. Byrne, B. Shen, and W. Wallace, "The Economics of Sustainable Energy for Rural Development: A Study of Renewable Energy in Rural China," *Energy Policy* 26 (1998): 45.
3. D. F. Barnes, K. Openshaw, K. R. Smith, and R. van der Plas, *What Makes People Cook with Improved Biomass Stoves?*, Technical Paper No. 242: Energy Series (Washington, D.C.: World Bank, 1994); and D. M. Kammen, "From Energy Efficiency to Social Utility: Improved Cookstoves and the *Small Is Beautiful* Model of Development," in J. Goldemberg and T. B. Johansson, eds., *Energy as an Instrument for Socio-Economic Development* (New York: United Nations Development Programme, 1995), 50.
4. S. Karekezi and T. Ranja, *Renewable Energy Technologies in Africa* (London: African Energy Policy Research Network and Stockholm Environment Institute and Zed Books, 1997); and D. Walubengo, "Commercialization of Improved Stoves: The Case of the Kenya Ceramic Jiko (KCJ)," in B. Westhoff and D. Germann, eds., *Stove Images: A Documentation of Improved and Traditional Stoves in Africa, Asia, and Latin America* (Brussels: Commission of the European Communities, 1995).
5. R. Duke and D. M. Kammen, "The Economics of Energy Market Transformation Programs," *Energy Journal* (forthcoming).
6. R. Acker and D. M. Kammen, "The Quiet (Energy) Revolution: The Diffusion of Photovoltaic Power Systems in Kenya," *Energy Policy* 24 (1996): 81; and M. Hankins, F. Omondi, and J. Scherpenzeel, *PV Electrification in Rural Kenya: A Survey of 410 Solar Home Systems in 12 Districts* (Nairobi: World Bank, 1997).
7. See K. L. Kozloff, "Rethinking Development Assistance for Renewable Electricity Sources," *Environment*, November 1995, 6; and Tata Energy Research Institute, *Climate Change: Post-Kyoto Perspectives from the South* (New Delhi, India, 1998).
8. On the scope of China's efforts in this area, see Chongqing Institute of Scientific and Technical Information, *New and Renewable Energy: Technologies and Products in China* (Chongqing, 1995); Y. Lu, *Fueling One Billion: An Insider's Story of Chinese Energy Policy Development* (Washington, D.C.: Washington Institute, 1993); and F. Dong, D. Lew, P. Li, D. M. Kammen, and R. Wilson, "Strategic Options for Reducing CO_2 in China: Improving Energy Efficiency and Using Alternatives to Fossil Fuels," in M. B. McElroy, C. P. Nielsen, and P. Leiden, eds., *Energizing China: Reconciling Environmental Protection and Economic Growth* (Cambridge, Mass.: Harvard University Press, 1998), 119. The latter volume offers a useful, coordinated effort to understand the lessons for energy technology and policy in China.
9. See Lu, note 8 above.
10. For more information on the Village Power conference, visit its Web site, *http://www.rsvp.nrel.gov/rsvp/tour/VPConference/vp98proceedings.html.*
11. M. R. Dove and D. M. Kammen, "The Epistemology of Sustainable Resource Use: Managing Forest Products, Swiddens, and High-Yielding Variety Crops," *Human Organization* 56, no. 1 (1997): 91.
12. See Kammen and Dove, note 1 above.
13. Visit *http://www.repp.org.*
14. Visit *http://www.resvp.nrel.gov/rsvp/.*
15. See Barnes et al., note 3 above; A. Cabraal, M. Cosgrove-Davies, and L. Schaeffer, *Best Practices for Photovoltaic Household Electrification Programs: Lessons from Experiences in Selected Countries* (Washington, D.C.: World Bank,

1996); and D. Anderson, "Renewable Energy Technology and Policy for Development." *Annual Review of Energy and Environment* 22 (1997): 187. Despite the more restricted role of subsidies in recent years, it is important to recognize and learn from the successful ones. In Nepal, for instance, more than 48,000 household biogas systems have been installed since 1973 (one-half of them since 1995). The cost of these systems ranges from $279 to $662. Dutch donors and the Nepal Agricultural Development Bank provide subsidies ranging from $103 in the central valley to $176 in remote mountainous regions. The amount of the subsidy is also larger (in percentage terms) for small systems. These subsidies have worked well due to careful planning and implementation. A single, well-tested design is used throughout the country, and it is installed by more than 40 private institutions. The subsidies are paid incrementally over three years to ensure sustained operation, and they have been used to leverage the development of high-quality digesters and competition among suppliers.

16. Kammen, note 3 above; and Walubengo, note 4 above.
17. K. R. Smith, G. Shuhua, K. Kun, and Q. Daxiong, "100 Million Biomass Stoves in China: How Was It Done?" *World Development* 18 (1993): 941.
18. Duke and Kammen, note 5 above; and S. Ryder, "The Poland Efficient Lighting Project." Contribution to the *Special Report on Methodological and Technological Issues in Technology Transfer* by Working Group II of the Intergovernmental Panel on Climate Change, 1999.
19. K. V. Ramani, *Rural Electrification and Rural Development: Rural Electrification Guidebook for Asia and the Pacific* (Bangkok: Asian Institute of Technology; and Brussels: Commission of the European Communities, 1992). Import tariffs are often a significant barrier to the dissemination of small-scale and decentralized energy technologies. See Acker and Kammen, note 6 above. Prices in different locations can also vary dramatically for other reasons. A recent study found that the cost of a solar home system ranged from $7 per watt in parts of China to more than $30 per watt in parts of Kenya and Sri Lanka (where the systems were sold as part of a tied-aid procurement program). Similar variations in price have been documented within China, Kenya, and Indonesia. See Cabraal et al., note 15 above.
20. J. A. Cohen and R. G. Noll, *The Technology Pork Barrel*, (Washington, D.C.: Brookings Institution, 1991).
21. For representative experience curves for various energy systems in the United States and Japan, see A. McDonald, "Combating Acid Deposition and Climate Change: Priorities for Asia," *Environment*, April 1999. Figure 1 on page 9.
22. For a formal treatment, see M. Spence, "The Learning Curve and Competition," Bell Journal of Economics 12 (1981): 49.
23. The analytic solution—though not the real-world one—would be to subsidize a single producer so that it would produce to the point where the price of the product equals the true (long-run) marginal cost of production. This is clearly not possible (or desirable) in either developed or developing countries. See Spence, note 22 above.
24. See Duke and Kammen, note 5 above.
25. The price elasticity of demand is defined as the percent change in demand stemming from a 1.0 percent reduction in price. An (absolute) value of 1.0 thus implies that demand will increase by the same percentage as the price declines, so that overall revenues (calculated as price times quantity) remain unchanged. Values higher than 1.0 imply an increase in revenues, values less than 1.0 a decrease.
26. See President's Committee of Advisors on Science and Technology Panel on U.S. Government Roles in International Cooperation on Energy Research, Development, Demonstration, and Deployment (ERD3), 1997; and Duke and Kammen, note 5 above.

27. Kozloff, note 7 above; and C. Macilwain, "World Bank Backs Third World Centres of Excellence Plan." *Nature* 396 (24/31 December 1998): 711.

The Commercial Dissemination of Photovoltaic Systems in Kenya

There has been an active commercial market for solar photovoltaic (PV) home systems in Kenya for more than 10 years. Some 80,000 systems have been sold, providing power to more than 1 percent of the rural population of 20 million people. The total installed capacity is in excess of 2 peak megawatts, with typical individual systems ranging from 20 to 40 peak watts and costs ranging from $300 to $1,500 per system. By contrast, the national Rural Electrification Program has connected less than 2 percent of rural households to the power grid. The Kenyan PV market evolved without subsidies or significant support from the government or multinational agencies, although several private and volunteer organizations were instrumental in disseminating information and offering training opportunities. Today, dozens of companies are active in the PV industry in East Africa, and annual sales exceed 20,000 systems and 300 kilowatts of capacity.[1] The success (and future potential) of the Kenyan market has attracted widespread international interest, and Kenya has been granted a World Bank Photovoltaic Market Transformation Initiative loan of more than $5 million.[2]

Until the late 1980s, PV electrification in East Africa was largely confined to systems owned by the very affluent or furnished by international donors to power wells, bore holes, schools, refrigerators in rural health clinics, and missions.[3] These systems generally arrived as complete packages, often imported directly from overseas via a local agent, distributor, or development group. Falling prices for crystalline and amorphous panels, increased awareness of the services that photovoltaics can provide, and recognition of the potential to commercialize PV systems all increased activity in the local energy market.

The most important aspect of the dissemination process in Kenya is the fact that it has been largely market driven. Initially, most systems were relatively large and possessed by rural elites. Although the income of system owners has historically been higher than average in rural areas, a recent survey of more than 400 households found that while their mean income was $108 per month, 75 percent earned less than $100 per month.[4] In accordance with the drop in the average income of system owners, the average system size has decreased from more than 40 peak watts to about 25 (including large numbers of 12 watt systems).

Growing interest in solar home systems in the late 1980s and early 1990s was accompanied by expansion of the retail network, with a diverse array of assembly, sales, installation, and maintenance businesses emerging. While solar panels are still imported, local companies now manufacture batteries for use in PV systems, and the "balance of system" (i.e., the electronics, charge controllers, lights, outlets, and other components) are either assembled or manufactured in Kenya. Local agents now sell more than one-half of all modules and three-fourths of the PV batteries. Sales of solar home systems have been increasing 10 to 18 percent per year, and this trend is expected to continue (along with further growth and activity in the commercial sector).

Photovoltaic systems have had a dramatic impact in Kenya. Surveys indicate that 60–90 percent of these systems have been performing well, with dead batteries accounting for the majority of those that were not in use. According to these surveys, most users purchased their systems for light, televison, and radio; they are pleased with them and would recommend them to others. In many locations, PV systems are cheaper than kerosene, diesel, and other energy alternatives, and in virtually every case the service they provide is superior in quality and reliability.

Kenya's experience offers an important model for other countries,[5] and Kenya and a number of other developing nations are now the target for international aid to expand the PV market in rural areas. Kenya has also become the focus of a regional PV market that extends into neighboring nations and other regions of Africa.

A variety of lessons emerges from the Kenyan experience. First, a relatively small number of organizations and individuals can provide the training and support services necessary for the success of an emerging technology. Second, subsidies are not necessarily needed to promote technology transfer, although logistical support, training, and performance standards all require policy attention and commitment of resources. Third, the diversity of commercial interests has been critical to the sustainability of the PV industry. Government and international policies can also be significant, especially those that promote independent power producers, limit or remove taxes and tariffs on clean energy alternatives, and provide credit or financing to both companies and end users.

Notes

1. M. Msinga, M. Harkins, D. Hirsch, and J. de Schutter, *Kenya Photo-Voltaic Rural Energy Project (KENPRED): Results of the 1997 Market Survey* (Nairobi: Energy Alternatives Africa, 1997).
2. R. Duke and D. M. Kammen. "The Economics of Energy Market Transformation Programs," *Energy Journal*, forthcoming.
3. R. Acker and D. M. Kammen, "The Quiet (Energy) Revolution: The Diffusion of Photovoltaic Power Systems in Kenya," *Energy Policy* 24 (1996): 81.
4. Ibid.; and Msinga et al., note 1 above.
5. A. Cabraal, M. Cosgrove-Davies, and L. Schaeffer, *Best Practices for Photovoltaic Household Electrification Programs: Lessons from Experiences in Selected Countries* (Washington, D.C.: World Bank, 1996).

Appendix

Milestones in the History of Energy & Its Uses

Energy has a long history. Beginning back before people could read and write, fire was discovered to be good for cooking, heating, and scaring wild animals away. Fire was possibly civilization's first great energy invention, and wood was the main fuel for a long time.

Wood

Pre-1885: Wood was the primary source for cooking, warmth, light, trains and steamboats. Cutting wood was time consuming, hard work. A common and meaningful poem was "Treat your wife fair and good, / Keep her stocked with firewood."

Electricity

1830–1839: Michael Faraday built an induction dynamo based on principles of electromagnetism, induction, generation and transmission.

1860s: Mathematical theory of electromagnetic fields published. Maxwell created a new era of physics when he unified magnetism, electricity and light. One of the most significant events, possibly the very most significant event, of the 19th century was Maxwell's discovery of the four laws of electrodynamics ("Maxwell's Equations"). This led to electric power, radios, and television.

Coal

1885–1950: Coal was the most important fuel. One half ton of coal produced as much energy as 2 tons of wood and at half the cost. But it was hard to stay clean in houses heated with coal.

Late 1860s: The steel industry gave coal a big boost.

1982: Coal accounted for more than half of the supply of electricity but little was used in homes. In terms of national electricity generation, hydro-power, natural gas, and nuclear energy contributed between 10 and 15 percent each.

Oil

By 1870: Oil had become the country's second biggest export.

1951–present: Oil has given us most of our energy. Automobiles increased the demand for oil.

1981: The U.S. Government decontrolled prices on petroleum and natural gas.

Nuclear

1906: Special theory of relativity written. Einstein created a new era of physics when he unified mass, energy, magnetism, electricity, and light. One of the most significant events, if not the very most significant event, of the 20th century was Einstein's writing the formula of $e=mc^2$: energy = mass times the square of the speed of light. This led to nuclear medicine—and a much longer life span, astrophysics, and commercial nuclear electric power.

1942: Scientists produced nuclear energy in a sustained nuclear reaction.

1957: The first commercial nuclear power plant began operating.

1995: Nuclear power contributed about 20 percent of the nation's electricity.

Transportation

1781: The stagecoach was the worldwide standard for passenger travel.

1800: Transportation as we know today was almost non-existent. Railroads covered far less territory. Trains were much smaller. Horse-drawn carts moved food and all other items on land, and barges moved them on rivers.

1881: The steam-powered railway train had become the worldwide standard for passenger travel.

1908: Henry Ford produced the Model T car. (Note that the Model T had been designed to use ethanol, gasoline, or any combination of the two fuels.)

1920: The Ford Motor Company manufactured the Model T in large numbers.

1949–2000: In transportation, use of energy is overwhelmingly petroleum. Energy for this use more than tripled from 1949 to 2000, with motor gasoline accounting for about two-thirds of it. Distillate fuel oil and jet fuel are other important petroleum products used in transportation.

1950–present: The National Highway Defense System opened interstate highways for fast trucks.

Energy Uses Have Changed

1800: The residential sector consumed most of America's energy.

1850–1980: The average energy that each person used increased steadily.

1979–1982:

- Energy consumption decreased ten percent.
- The industrial sector cut its consumption by 20 percent.
- The residential and commercial sectors energy consumption stayed about the same.

1950:

- Distillate fuel oil heated about 22 percent of U.S. households.
- Over a third of all U.S. housing units were warmed by coal.
- Natural gas was used to warm about 25 percent of U.S. households.
- Electricity was used to warm only 0.6 percent of U.S. households.

1978: Microwave ovens were located in 8 percent of U.S. households.

1990: 16 percent of households owned one or more personal computers.

1997:

- Only about 11 percent of all U.S. housing units were warmed by distillate fuel oil.
- Only 0.2 percent of all U.S. housing units were warmed by coal.
- More than 50 percent of all U.S. households used natural gas for warmth.
- Electricity was used in 29 percent of U.S. households for warmth.
- 35 percent of U.S. households had personal computers.
- Microwaves could be found in 83 percent of U.S. households.
- 99 percent of U.S. households had a color television.
- 47 percent had central air conditioning.
- 85 percent of all households had one refrigerator, the remaining 15 percent had two or more.

Energy Consumption in the United States, 1775–1999

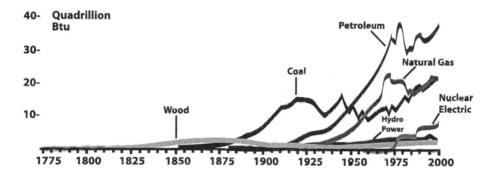

General Milestones

Energy Consumption Milestones

1850–1972: Energy use per person increased over time.

1972–1982: Per capita energy consumption decreased ten percent.

1992: Thirty four percent of total energy was used to make electricity.

General Energy Milestones

Beginning in Time Immemorial—Fire is discovered to be good for cooking, heating and scaring wild animals away.

3000 B.C.: Humans began using petroleum. Mesopotamians of that era used *rock oil* in architectural adhesives, ship caulks, medicines, and roads. The Chinese of two millennia ago refined crude oil for use in lamps and in heating homes.

2800 B.C.: Sumerian clay tablets record oil sales by amphorae. This olive oil is used in Sumerian lamps and in cooking.

1100 B.C.: Scattered records exist of the use of coal as a fuel.

200 B.C.: First practical use of natural gas is attributed to the Chinese. They used it to make salt from brine in gas-fired evaporators, boring shallow wells with crude percussion rigs and conveying the gas to the evaporators via bamboo pipes.

250–400 A.D.: Near Arles in France, the Romans built a water-powered mill with 16 wheels that had an output of over 40 horsepower.

600 A.D.: Arab and Persian chemists discovered that petroleum's lighter elements could be mixed with quicklime to make *Greek fire*, the napalm of its day.

800–1500 A.D.: Viking ships use wind energy to dominate the oceans.

874 A.D.: Iceland received its first inhabitants from Norway. Geothermal energy keeps them warm.

1400s: The invention of firebricks, which made chimneys cheap to build, helped create a home heating market for coal.

1570s: Despite its drawbacks (smoke and fumes), coal was firmly established as a domestic fuel. The total amount of coal consumed in the United States in all the years before 1800 was an estimated 108,000 tons, much of it imported. The U.S. market for coal expanded slowly and it was not until 1885 that the young and heavily forested nation burned more coal than wood.

1700s:

- A French military officer noted that Indians living near Fort Duquesne (now the site of Pittsburgh) set fire to an oil slicked creek as part of a religious ceremony.
- Nearly all energy in the United States was still supplied by muscle power and fuel wood with small amounts of coal used to make coke—vital for casting the cannon. Mills made use of waterpower, and wind power enabled transport by ship.

1800–1826:

- First electric battery made.
- In England, the Domesday Book survey of 1806 counted 5,624 mills in the South and East (found throughout Europe and elsewhere) and were used for a wide variety of mechanical tasks in addition to milling, from pressing oil to making wire. Some installations were surprisingly large.

- Humphrey Davy built a battery-powered arc lamp.
- The first energy utility in U.S. was founded.
- The relationship between electricity and magnetism was confirmed.
- First electric motor developed by Faraday.
- Ohms Law published.

1830–1839:

- Michael Faraday built an induction dynamo based on principles of electro-magnetism, induction, generation and transmission.
- First industrial electric motors built.
- First fuel cell designed.

1840–1865:

- Experiments were conducted with battery powered electric trains.
- A giant 72-foot waterwheel with an output of 572 horsepower, dubbed Lady Isabella, was erected at a mine site on the Isle of Mann.
- The first oil well, Edwin L. Drake's homemade drilling rig at Titusville, Pennsylvania, struck oil on August 27, 1859.
- Mathematical theory of electromagnetic fields published. Maxwell created a new era of physics when he unified magnetism, electricity and light. One of the most, if not the very most, significant event of the 19th century was Maxwell's discovery of the four laws of electrodynamics. This led to electric power, radios, and television.

1870–1880:

- Draft animals accounted for more than half of the total horsepower of all prime movers.
- Gas turbine patented.
- First combustion engine designed to use alcohol and gasoline was made.
- Edison Electric Light Co. (U.S.) and American Electric and Illuminating (Canada) founded.
- First commercial power station opens in San Francisco using brush generator and arc lights.

1881–1887:

- Thomas Edison opened the first electricity generating plant (in London) in January 1881.
- Edison's Pearl Street Station opened in New York as the first American plant to generate electricity. Within a month after the first electricity generating plant in America began operating, it was feeding 1,300 lightbulbs. Within a year, it was feeding 11,000 bulbs—each a hundred times brighter than a candle.

- First hydroelectric station opens (Wisconsin).
- Transformer invented.
- Steam turbine invented.
- Stanley develops transformer and invents the alternating current electric system.
- Tesla invented the induction motor with a rotating magnetic field. This made unit drives for machines and AC power transmission economically feasible.
- Electron discovered.

1898–1988: Energy recovery from garbage incineration started in New York City. The primary focus for the next eight decades was on waste volume reduction through incineration, and energy recovery was used primarily for process heat.

1900s–1950: Windmills were used to pump water and to generate electricity at remote sites.

1900–1910:

- Highest voltage transmission line (60 Kilovolt) built.
- Texas's vast Spindletop Oil Field was discovered.
- 5-Megawatt turbine was built for Fisk St. Station (Chicago).
- First successful gas turbine built (France).
- World's first all turbine station begins operation (Chicago).
- Shawinigan Water & Power installs world's largest generator (5,000 Watts) and world's longest and highest voltage line (136 Km and 50 Kilovolts) to Montreal.
- A 5 megawatt steam-driven turbine generator (the first of its type and the largest of any generator then built) was commissioned.
- First geothermal electricity commercialization began in Italy.
- First electric vacuum cleaner produced.
- First electric washing machine sold.
- Henry Ford's Model T designed to use ethanol, gasoline, or any combination of the two fuels.
- First pumped storage plant (Switzerland).
- Special theory of relativity published. Einstein created a new era of physics when he unified mass, energy, magnetism, electricity and light. One of the most, if not the very most, significant event of the 20th century was Einstein's discovery of $e=mc^2$. This led to nuclear power, nuclear weapons, nuclear medicine and astrophysics.

1911–1919:

- First air conditioning unit built.

- First air pollution control device (a cinder catcher was installed).
- First electric refrigerator made.
- Federal government begins construction of Muscle Shoals Dam (origin of Tennessee Valley Authority).
- End of World War I (Nov. 1918).
- Atomic fission theory developed.

1920–1930:

- First U.S. station to only burn pulverized coal.
- Federal Power Commission (FPC) created.
- Coal accounted for about 75 percent of U.S. total energy use.
- Connecticut Valley Power Exchange (CONVEX) starts, pioneering interconnection between utilities.
- First television components are built.
- First all-welded natural gas pipeline over 200 miles in length was built from Louisiana to Texas.
- Construction of Boulder Dam began.
- Federal Trade Commission begins investigation of holding companies.
- Jet engine patented.
- Kerosene and fuel oil began displacing wood for some commercial, transportation, and residential uses.

1933–1939:

- Tennessee Valley Authority (TVA) established.
- Nuclear chain reaction described.
- Public Utility Holding Company Act.
- Federal Power Act.
- Securities and Exchange Commission.
- Bonneville Power Administration.
- First night baseball game in major leagues.
- President Franklin D. Roosevelt signed into law the Rural Electrification Administration (REA) Act. REA loaned money at low interest and helped to set up electricity cooperatives.
- Highest steam temperature in electricity generation reaches 900 degrees Fahrenheit vs. 600 degrees Fahrenheit.
- 287 Kilovolt line runs 266 miles to Boulder (Hoover) Dam.
- Boulder Dam completed.
- Man-made fission of uranium occurs.
- First jet flight engine developed.

1940–1949:

- First U.S. fuel ethanol plant built.
- First sustained nuclear reaction happened in Chicago. Enrico Fermi's group achieved the reaction for the first time at the University of Chicago in a primitive graphite moderated reactor built on a vacant squash court.
- First atomic bomb detonated.
- Atomic Energy Act establishes Atomic Energy Commission (AEC).
- Transistor invented.

1950s:

- Electricity and natural gas displaced wood heat in homes and commercial buildings.
- Distillate fuel oil heated about 22 percent of U.S. households.
- Over a third of all U.S. housing units were warmed by coal.
- Natural gas was used to warm about 25 percent of U.S. households.
- Electricity was used to warm only 0.6 percent of U.S. households.
- Most coal was consumed in the industrial sector, but many homes were still heated by coal and the transportation sector still consumed significant amounts in steam-driven trains and ships.
- An experimental reactor sponsored by the U.S. Atomic Energy Commission generated the first electricity from nuclear power. The British completed the first operable commercial reactor, at Calder Hall, in 1956. The U.S. Shippingport unit, a design based on power plants used in nuclear submarines, followed a year later.
- First practical nuclear reactor for submarines.
- First 345 Kilovolt transmission line.
- First nuclear power station ordered.
- First high voltage direct current (HVDC) line (20 megawatts/1900 Kilovolts, 96 Km).
- Atomic Energy Act of 1954 allows private ownership of nuclear reactors.
- Nuclear submarine (*Nautilus*) commissioned by U.S. Navy.

1960s:

- Clean Air Act.
- Northeast electricity blackout.
- National Electric Reliability Council (NERC) formed.
- National Environmental Policy Act of 1969.

1970s:

- Environmental Protection Agency (EPA) formed.
- Water and Environmental Quality Act.

- The Clean Air Act (CAA) Amendments significantly expanded the role of the Federal Government in controlling air pollution.
- The Clean Water Act of 1972 signed into law, affecting discharges from power plants and energy-intensive industries.
- The Arab Oil Embargo was instituted against the United States and Holland by several Arab nations.
- To protect against supply disruptions, the United States began to build a Strategic Petroleum Reserve.
- The Energy Policy and Conservation Act (EPCA) of 1975 was enacted to achieve a number of goals, among which was the establishment of the Strategic Petroleum Reserve and increasing automobile fuel efficiency.
- Brown's Ferry nuclear accident.
- New York City electricity blackout.
- Department of Energy (DOE) formed.
- Carter energy plan developed.
- The National Renewable Energy Laboratory (NREL) was formed as a national laboratory that provides research and development support for solar, photovoltaic, geothermal, wind and other renewable technologies.
- Public Utilities Regulatory Policies Act (PURPA) passed, ends utility monopoly over generation.
- Natural Gas Policy Act partially deregulates wellhead prices.
- Power Plant and Industrial Fuel Use Act limits use of natural gas in electric generation (repealed 1987).
- Microwave ovens located in 8 percent of U.S. households.
- Three Mile Island nuclear accident happened.

1973–1993: Fuel consumption per motor vehicle fell, miles traveled per vehicle generally fell until the early 1980s and then resumed a pattern of increase, and the fuel rate (i.e., miles per gallon) improved greatly.

1980s:
- First U.S. wind farm built.
- Pacific Northwest Electric Power Planning and Conservation Act establishes regional regulation and planning.
- PURPA ruled unconstitutional by Federal judge.
- U.S. Supreme Court upholds legality of PURPA in *Federal Energy Regulatory Commission v. Mississippi*.
- Hydrothermal generating capacity in U.S. exceeds 1,000 megawatts.
- Nova Scotia tidal power plant is the first of its kind in North America.
- Citizens Power, first power marketer, goes into business.
- Chernobyl nuclear accident (USSR).

- Washington had recycling legislation mandating a state Municipal Solid Waste reduction goal through recycling.

1990s:

- Clean Air Act amendments mandate additional pollution controls.
- More than 2,200 megawatts of wind energy capacity was installed in California—more than half of the world's capacity at the time.
- 16 percent of households owned one or more personal computers.
- National Energy Policy Act passed.
- 15 states had adopted recycling legislation. Today, there are over 7,000 recycling programs, covering one-third of the U.S. population; more than 1,000 bills are proposed per year that endorse some type of recycling law, incentive, or program. Many state goals are to reduce Municipal Solid Waste (MSW) by 50 percent or more.
- Commercial production of variable speed wind turbine begins (U.S.).
- Worldwide geothermal capacity exceeds 6,000 megawatts.
- ISO New England begins operation (first Independent System Operator).
- New England Electric sells power plants (first major plant divestiture).
- 99 percent of U.S. households had a color television and 47 percent had central air conditioning.
- Eighty-five percent of all households had one refrigerator; the remaining 15 percent had two or more.
- Microwaves could be found in 83 percent of U.S. households.
- 35 percent of U.S. households had personal computers.
- Only 0.2 percent of all U.S. housing units were warmed by coal.
- Only about 11 percent of all U.S. housing units were warmed by distillate fuel oil.
- More than 50 percent of all U.S. households used natural gas for warmth.
- Electricity was used for warmth by 29 percent of U.S. households.
- California opened deregulated electricity market.
- Scottish Power (U.K.) bought Pacificorp in the first foreign takeover of a U.S. utility. National (U.K.) Grid then announced purchase of the New England Electric System.
- Electricity marketed on Internet for the first time.
- Federal Energy Regulatory Commission (FERC) issued Order 2000, promoting regional transmission.

2000: Fuel cells achieved new highs in efficiency.

Source: *Energy Information Administration, www.eia.doe.gov*

Bibliography

Books

Asmus, Peter. *Reaping the Wind: How Mechanical Wizards, Visionaries, and Profiteers Helped Shape Our Energy Future*. Washington, D.C.: Island Press, 2000.

Bent, Robert, Randall Baker, and Lloyd Orr, eds. *Energy: Science, Policy, and the Pursuit of Sustainability*. Washington, D.C.: Island Press, 2002.

Berinstein, Paula. *Alternative Energy: Facts, Statistics, and Issues*. Westport, CT: Onyx Press, 2001.

Borowitz, Sidney. *Farewell Fossil Fuels: Reviewing America's Energy Policy*. New York: Plenum Publishing, 1998.

Burleson, Clyde W. *Deep Challenge!: The True Epic Story of Our Quest for Energy Beneath the Sea*. Houston, TX: Gulf Professional Publishing Company, 1998.

Casten, Thomas R. *Turning Off the Heat: Why America Must Double Energy Efficiency to Save Money and Reduce Global Warming*. Amherst, NY: Prometheus Books, 1998.

Christensen, John. *Global Science: Energy, Resources, Environment*. Dubuque, IA: Kendall/Hunt Publishing, 1994.

Clark, Wilson, and Jake Page. *Energy, Vulnerability, and War: Alternatives for America*. New York: Norton, 1981.

Cozic, Charles P., and Matthew Polesetsky. *Energy Alternatives*. San Diego, CA: Greenhaven Press, 1992.

Dunn, Seth. *Micropower: The Next Electrical Era*. Washington, D.C.: World Watch Institute, 2000.

Economides, Michael, Ronald Oligney, and Armando Izquierdo. *The Color of Oil: The History, the Money and the Politics of the World's Biggest Business*. Houston, TX: Round Oak Publishing, 2000.

Garwin, Richard L., and Georges Charpak. *Megawatts and Megatons: A Turning Point in the Nuclear Age*. New York: Knopf, 2001.

Gelbspan, Ross. *The Heat Is On: The High Stakes Battle over Earth's Threatened Climate*. Cambridge, MA: Perseus Press, 1997.

Gipe, Paul. *Wind Energy Comes of Age*. New York: John Wiley & Sons, 1995.

Hoffman, Peter. *Tomorrow's Energy: Hydrogen, Fuel Cells, and the Prospects for a Cleaner Planet*. Cambridge, MA: MIT Press, 2002.

Karl, Terry Lynn. *The Paradox of Plenty: Oil Booms and Petro-States*. Berkeley, CA: University of California Press, 1997.

Koppel, Tom. *Powering the Future: The Ballard Fuel Cell and the Race to Change the World*. New York: John Wiley & Sons, 1999.

Nakicenovic, Nebojsa, Arnulf Grübler, and Alan McDonald, eds. *Global Energy Perspectives*. New York: Cambridge University Press, 1998.

Nansen, Ralph. *Sun Power: The Global Solution for the Coming Energy Crisis.* Melbourne, Australia: Ocean Press, 1995.

Pachauri, Rajendra K. *The Political Economy of Global Energy.* Baltimore, MD: John Hopkins University Press, 1985.

Rashid, Ahmed. *Taliban: Militant Islam, Oil and Fundamentalism in Central Asia.* New Haven, CT: Yale University Press, 2000.

Rhee, Kee H., ed. *Clean Use of Coal.* Collingdale, PA: DIANE Publishing, 1995.

Rapporteur, Paul Runci, and Roger W. Sant. *After Kyoto: Are There Rational Pathways to a Sustainable Global Energy System?* Queenstown, MD: Aspen Institute, 1999.

Van Der Leeuw, Charles. *Oil and Gas in the Caucasus & Caspian: A History.* New York: Palgrave, 2000.

Wasserman, Harvey. *The Last Energy War: The Battle over Utility Deregulation.* New York: Seven Stories Press, 2000.

Web Sites

For those who wish to find more information online about energy, this section lists various Web sites that may be of interest. These sites are only a small fraction of the many sites on the subject, but we hope they will serve as a starting point. Due to the nature of the Internet, the continued existence of a site is never guaranteed, but at the time of this book's publication, all of these Internet addresses were in operation.

Alliance to Save Energy

www.ase.org

The Alliance to Save Energy, a nonprofit advocacy group, provides on-line materials to educators and consumers about reducing energy use and increasing efficiency.

Climate Action Network

www.climatenetwork.org

The online home of a coalition of non-governmental organizations, this site discusses the issues surrounding the Kyoto Protocol, including nuclear energy, emissions trading, and compliance. There are also reports about energy and climate from various members.

Energy Information Administration

www.eia.doe.gov

The official site of the Energy Information Administration provides extensive information about energy use in the U.S. and abroad. The information is organized by fuel type, region, price, and sector. There are also research materials appropriate for high school classroom use.

Energy Trends

energytrends.pnl.gov

This site posts reports from the Pacific Northwest National Laboratory, under the U.S. Department of Energy. The reports analyze energy trends and policy in the major industrialized countries.

Environmental Protection Agency

www.epa.gov

The official site of the Environmental Protection Agency provides information about global climate change.

Green Mountain Energy Company

www.greenmountain.com

The Web site of the Green Mountain Energy Company, the largest provider of less-polluting electricity generated from renewable energy sources, posts news about renewable energy use in the U.S.

Hydrogen and Fuel Cell Newsletter

www.hfcletter.com

This electronic newsletter, founded in 1986, covers events in the emerging fields of fuel cell and hydrogen technology. The Web site posts each month's feature stories.

IEA Coal Research Centre

www.iea-coal.org.uk

This Web site provides information about clean-coal technology and production.

Nuclear Energy Institute

www.nei.org

The Nuclear Energy Institute, an industry policy group, posts data and news about nuclear energy on their site, as well as materials for classroom use.

U.S. Department of Energy

www.energy.gov

The official site of the Department of Energy posts news about departmental activities and includes a comprehensive research database aimed at students.

World Energy Council

www.worldenergy.org

The World Energy Council is an umbrella organization of energy producers, policy-makers, and environmental groups. Their site provides data about energy production, consumption, and emissions.

Additional Periodical Articles with Abstracts

Those interested in learning more about energy issues may refer to the following list of articles. Readers who require a more comprehensive selection are advised to consult *Reader's Guide Abstracts* and other H.W. Wilson indexes.

The Oil Spigot May Be Closing: Views of C. Maxwell. Robert Barker. *Business Week* p160 November 12, 2001

Charles Maxwell, senior energy analyst for institutional broker Weeden and Co., predicts that oil prices by 2006 or so will be sharply and permanently higher. By summer 2002, when Maxwell foresees that the economy and energy use will pick up, oil prices will also rise. Although not every energy analyst agrees with his forecast, even if he is partly right, the implications for investors would be huge. The social dislocations and political turmoil that most concern Maxwell are examined.

What's Greener and Has Legs? Bush's Energy Plan. Laura Cohn. *Business Week* p47 July 23, 2001

According to Cohn, George W. Bush's energy plan is far from dead. The President's plans to increase drilling on federal lands, expand nuclear power, soften environmental protections, and heighten refinery capacity were panned by the public and marked the abrupt end of his six-month honeymoon in office. Three unexpected events have converged to change the landscape dramatically, however. First, Bush responded to falling polls by shifting his rhetoric from boosting the energy supply to energy efficiency and conservation, placing him on the same page as congressional Democrats and GOP moderates. Second, the declining prices of natural gas and gasoline in recent weeks lessened the crisis atmosphere on Capitol Hill and prompted both sides to ease off the blame game. Third, the Democratic takeover of the Senate means that business-friendly consensus-builder Jeff Bingaman replaces pro-oil stalwart Frank H. Murkowski as head of the Senate Energy Committee.

Energy Conservation: An Idea Whose Time Has Come Again. Robert Kuttner. *Business Week* p26 November 19, 2001

Kuttner writes that the next major economic blow could easily be an energy crisis. Should the war in Afghanistan drag on or expand beyond the Afghan borders, the Middle East could become less stable and with it the short-run supply of oil. A far more serious problem is the dwindling of worldwide oil reserves. In a new book, *Hubert's Peak: The Impending World Oil Shortage*, oil geologist Kenneth S. Deffeyes has predicted that world oil production will peak sometime in this decade and will slowly and irreversibly decline from then on. Short-term events and long-term prospects cry out for an energy policy that emphasizes conservation and a shift to renewable sources of power. Unfortunately, Kuttner says, the Bush administration is following a policy that stresses new drilling.

Power Switch: Move Toward Solar and Wind Power in Canada.
Lawrence Scanlan. *Canadian Geographic* v. 121 pp54–62 May/June 2001

Scanlan explains that, although the move toward solar and wind power has
not exactly been revolutionary, it is growing in momentum. Wind turbines are
the fastest-growing energy source in the world, and experts are forecasting
that pollution-free renewables will represent a significant share of the energy
market by 2030. Renewable energy has long promised a bright future just
around the corner. The rub has always been expense and reliability; Canadi-
ans like our power inexpensive and simple and balk at paying more for clean
energy. Traditional energy sources have never been more expensive or uncertain, however, and now going green, which has always made environmental
sense, could soon make economic sense as well.

Clean Coal Back on Front Burner. Jeff Johnson. *Chemical and Engineering News* v. 79 pp37–40 May 7, 2001

The Department of Energy's 15-year-old Clean Coal Technology Program has
been awarded a $2 billion infusion of funds, spread over ten years, by the
Bush administration. The program will receive its first instalment of $150
million in the 2002 budget proposal. This program is discussed in detail, and a
sidebar examines the support that the coal and energy industries gave to
President Bush's presidential election campaign.

What to Do about the Energy Crunch. Irwin M. Stelzer. *Commentary* v.
111 pp33–7 March 2001

Stelzer asserts that Americans will have to look to themselves rather than the
Organization of Petroleum Exporting Countries (OPEC) to find a solution to
the country's energy problems. Despite the fact that America accounts for
about 25 percent of the world's oil consumption, the nation produces only
about 10 percent of the world's oil and owns less than 3 percent of global
untapped oil reserves. By acting together, the 12 members of OPEC can significantly affect the economic well-being of every society that depends on oil for
power and heating. Consequently, Stelzer writes, there needs to be a more
realistic assessment of the advantages of increased domestic production mea-
sured against its environmental costs, and steps must be taken to weaken the
OPEC cartel.

Lovin' Hydrogen. Brad Lemley. *Discover* v. 22 pp52–57, 86 November 2001

According to energy visionary Amory Lovins, a profitable, pollution-free
hydrogen-based economy will come about in the near future. Hydrogen contains more chemical energy per unit mass than any other known fuel, and
most modern hydrogen-based schemes involve a fuel cell, which combines
hydrogen with oxygen to produce electricity. Unlike other environmentalists,
who contend that clean technologies will flourish only with government assistance, Lovins claims that the entire fossil fuel-based economy will give way to
hydrogen because of simple obsolescence and efficiencies. His proposal for converting to hydrogen concentrates on the money that can be made, and he tries
to encourage corporations and governments along the hydrogen route by con-

sulting, spinning off companies, and preaching the benefits of hydrogen. However, many energy experts contend that hydrogen will continue to play a marginal role for decades due to the expense of making hydrogen, as well as the technical barriers that must be surmounted.

Invisible Energy: Solar Panels and Fuel Cells Incorporated into Four Times Square and Other Buildings. Kathryn Sergeant Brown. *Discover* v. 20 pp36+ October 1999

Brown writes that personal power generators are becoming increasingly popular. Local blackouts and Y2K fears have home owners and office managers worried about a power grid failure, while freestanding energy sources are getting cleaner, cheaper, and more attractive. Architects who once avoided bulky solar panels are amazed by solar cells incorporated discreetly into the surface of roofing shingles, windows, skylights, and wall facades. In the United States, President Clinton has introduced the Million Solar Roofs Initiative to place solar-energy tiles on the tops of 1 million buildings by 2010. Silent and clean, fuel cells create electricity by mixing oxygen with hydrogen that is usually taken from a fuel like natural gas or methanol, initiating chemical reactions that spark a current. Although Four Times Square, a new 48-story office tower in New York City, uses fuel cells for only a fraction of its total power needs, in other circumstances the cells could be more dominant. Wherever there is a source of hydrogen-rich gas, one might think of fuel cells, says Adam Serchuk of the Renewable Energy Policy Project in Washington, D.C.

Searching for Energy Efficiency on Campus: Clark University's 30-year Quest. Joseph F. DeCarolis, Robert L. Goble, and Christoph Hohenemser. *Environment* v. 42 pp8–20 May 2000

The ways in which Clark University in Worcester, Massachusetts, set about becoming energy efficient are discussed. The university's efforts were driven, above all, by responses to the external market for fossil fuels. In the early years of the 1970s oil crisis, the university tightened controls over heat and electric end use. Later, it was proposed that a grid-connected cogeneration plant be built and installed. An important element of these initiatives was the fact that teaching and learning were included in the decision-making process.

Bringing Power to the People: Promoting Appropriate Energy Technologies in the Developing World. Daniel Kammen. *Environment* v. 41 pp10–15+ June 1999

Kammen asserts that small-scale energy systems offer a viable alternative to traditional carbon-intensive energy sources that have high economic and environmental costs. The systems could prove to be both sustainable and beneficial to human health, unlike the biomass fuels presently used by more than 2 billion people worldwide. Small-scale energy systems are hampered by a lack of research and development, however, and the institutions, market policies, and training required to back them up and ensure their sustainable implementation are only starting to appear. The writer discusses some of the recent improvements in small-scale energy systems, the lack of institutional capacity for the development of such systems, the progress that has been made in developing countries, and a number of recommendations for promoting the systems and sustaining their markets.

Energy's Eastern Front. Benjamin Fulford. *Forbes* v. 168 pp60–2 December 24, 2001

The West is to invest $45 billion in oil and gas exploration on a desolate island off Russia. Oil giants such as Royal Dutch/Shell, Exxon-Mobil, and Chevron Texaco are preparing to turn Russia's far eastern Sakhalin into a huge oil-and-gas hub serving China, India, Korea, and other regions of Asia for at least the next half-century. The plans will require dozens of offshore platforms; liquefied natural gas plants; pipelines to China, Korea, and possibly Japan; and all the infrastructure to support these structures. The technical, logistical, and bureaucratic hurdles are enormous, but the writer argues that the great potential rewards make the Sakhalin gamble seem worthwhile. A sidebar outlines the obstacles between Sakhalin gas or oil and the Japanese.

Breaking OPEC's Grip. Nelson D. Schwartz. *Fortune* v. 144 pp78–88 November 12, 2001

Schwartz argues that, although America will continue to depend on imported oil, that does not mean that OPEC will always have the upper hand. Rigs in the Middle East can produce 20,000 barrels of oil a day, yet the United States still gets almost one-fifth of its domestic oil from trippers and other marginal wells dispersed from California to Pennsylvania. Energy independence is unrealistic, according to veteran petroleum expert Edward Morse, because America uses 25 percent of the world's oil but only has 3 percent of its reserves. Ironically, OPEC's success in keeping prices quite high has undercut its market share: simple economics dictates that those higher prices have stimulated exploration and development worldwide, especially in areas where oil is more expensive to produce. Non-OPEC oil will come from massive new discoveries in deepwater fields off Angola and Nigeria, in the Gulf of Mexico, and in the Caspian Sea, for starters.

Britain's Green Agenda. Tony Blair. *The Futurist*. v. 35 pp7–8 July/August 2001

In an article adapted from an address to the World Wildlife Fund Conference on Global Environmental Challenges on March 6, Britain's prime minister discusses his country's environmental commitment. Following the Kyoto process, Britain promised to reduce emissions by 12.5 percent, over twice the average commitment, and it has established a program that is projected to cut greenhouse gas emissions by 23 percent by 2010. Business, technology, and environmental protection must all be harnessed to promote renewable energy through green technologies that are about to become one of the next waves in the knowledge economy revolution. Britain is also leading efforts to transfer technologies that facilitate sustainable development to the developing world.

And Now a Word from Our Sponsor. *Harper's* v. 303 pp19–20 December 2001

A reprint of a testimony in a lawsuit filed in June against Exxon-Mobil for complicity in human rights abuses committed by Indonesian military forces, the Tentara Nasional Indonesia (TNI). The suit alleges that Exxon-Mobil provided barracks where victims were tortured and heavy equipment that was used to dig mass graves. International Labor Rights Fund, which has brought the cases, accuses the company of hiring TNI to protect its natural-gas operations in the Aceh province.

Springtime for Nuclear. Bill Mesler. *The Nation* v. 273 pp16–20 July 23/30, 2001

Messler writes that the nuclear industry has witnessed a dramatic turnaround in recent times. Nuclear power was once so politically untouchable that it spent the larger part of 20 years in duck-and-cover mode. It was pronounced dead not just by environmentalists but also by investors stung by its high costs and risks and fearful that the government subsidies that kept the industry afloat would eventually dry up. Nuclear plants are currently selling for record prices, however, and stock in the nation's largest nuclear holding company has doubled in value in the past year and a half. Furthermore, Messler asserts, the most pro-nuclear administration since that of Richard Nixon is in the White House, and according to a recent AP poll, 50 percent of all Americans support nuclear power.

Greenbacks: Businesses See Profits in the Kyoto Treaty. *National Review* v. 50 pp26+ December 21, 1998

The Kyoto Protocol, which if adopted by the United States would require dramatic cuts in the use of fossil fuels, is being viewed by some companies as an opportunity for profit. Alternative-energy companies and other associations in the Business Council for Sustainable Energy (BCSE) stand to benefit from stricter government controls on the emissions of greenhouse gases. Emission controls would force consumers to switch to cleaner and substantially more expensive alternatives. Other BCSE members would use emission controls to market energy-efficiency technologies and controls.

A Common-sense Solution to Global Warming. James Hansen. *New Perspectives Quarterly* v. 18 pp41–2 Winter 2001

Hansen writes that climate changes are likely to persist during the 21st century. More and more evidence that the world is slowly but surely getting warmer continues to accumulate. On the planet's surface, the average warming is now one degree Fahrenheit in the past 100 years. One worry is that, as glaciers melt, the sea level will rise. A rise of three feet, which is possible in this century, would submerge island nations such as the Maldives and the Marshall Islands. This would have devastating consequences in areas such as Bangladesh or the Nile Delta. Some temperature increase is unavoidable because it will take decades to phase in new energy infrastructure technologies that produce less carbon dioxide, the main cause of global warming. The most sensible approach will be to move forward by attacking air pollution, improving energy efficiency, and developing renewable energy sources.

Many Utilities Call Conserving Good Business. Timothy Egan. *New York Times* pA1 May 11, 2001

Egan explains how electric utilities in California and Washington State have turned to creative conservation programs, and their efforts have paid off. By promoting energy efficiency, the utilities have managed to avoid building additional power plants and, in the case of Los Angeles and Sacramento, have avoided rolling blackouts. Despite the Bush administration's shift away from conservation as part of its energy policy, energy conservation programs are here to stay because consumers want them. Conservation programs developed by Avista Utilities in Spokane, Wash., and city utilities in Los Angeles, Sacramento, and Seattle are described.

In the Oil-rich Nigeria Delta, Deep Poverty and Grim Fires. Norimitsu Onishi. *New York Times*, pA1 August 11, 2000

Because of Nigeria's complicated history of ethnic politics, residents of the Niger River delta have remained stunningly and desperately poor. The Niger River delta has been one of the world's top oil producing regions and has been the center of violent despair for years. An explosion on July 10 killed residents who were gathering fuel from a leaking pipe and then selling it on the black market. Nigeria leads the world in pipeline explosions, which are often the result of sabotage.

Alaska: Oil's Ground Zero. Jeffrey Bartholet. *Newsweek* v. 138 pp18–23 August 13, 2001

The debate is growing over whether to drill or preserve one of the last wild places of America, the Arctic National Wildlife Refuge (ANWR) in Alaska. Alaska has by far the largest area of wilderness lands in the United States, but it also contains the nation's two largest oil fields and is second only to Texas in proven reserves of crude. The Bush administration wants to drill in the ANWR coastal plain, arguing that new oil discoveries are crucial to preserve the lifestyles all Americans have become used to. Nobody actually knows how much oil is under the plain, however. Environmentalists argue that by increasing energy efficiency, Americans could save much more oil than could ever be drilled in places such as ANWR. Defenders of the wilderness also believe that the ecosystem has an intrinsic value beyond oil. The energy bill allowing Arctic refuge drilling has been approved by the House of Representatives, but the issue now moves to the Senate, where leading Democrats have vowed to keep oilmen out.

Alternative Energy Comes of Age. David Case. *Rolling Stone* pp39–41 September 13, 2001

Case argues that alternative energy has come of age. The four chief technology groups generating sustainable and environmentally friendly electricity—wind, solar, geothermal, and biomass—each use forces that will be around as long as the planet is: heat, motion, and gravity. Following over 30 years on the fringes, renewable energy is now big business. Leading corporations, mainly European ones, control vital sectors of the green-energy industry, and even some major oil firms have been far more vigorous in promoting alternative energy than the Bush administration. It is now clear that green energy helps people live longer by keeping the air cleaner, creates jobs in the United States, and eradicates the need to negotiate with dictators for oil. In addition, because green energy does not rely on far-flung fuel sources, ratepayers do not have to be concerned about fluctuating prices.

Bright Future—or Brief Flare—for Renewable Energy? Kathryn Sergeant Brown. *Science* v. 285 pp678–80 July 30, 1999

Brown writes that solar, wind, and other forms of renewable energy are making gains as some U.S. states deregulate their power markets. California, Pennsylvania, Texas, and at least 20 more states have deregulated or plan to deregulate their markets, giving renewable energy companies a chance to compete for the $250 billion a year in sales racked up by utilities. Currently, wind and solar power contribute less than 2 percent of the United States' total energy, even though global wind and solar power capacities in megawatts have been growing by around 11 percent and 16 percent a year, respectively, since 1990, according to the Worldwatch Institute. For renewable energy to succeed, however, it must become cheaper, and this will require enhanced technologies.

Fossil Fools: Need for Rational Energy Policy. Carl Pope. *Sierra* v. 85 pp14–16 July/August 2000

Pope asserts that it appears unlikely that the United States will ever adopt a reasonable energy policy. Neither George Bush nor Bill Clinton fulfilled their vows to decrease and stabilize the hydrocarbon emissions that lead to global warming because oil firms, car manufacturers, and others spent millions to persuade Americans that a reduction in fossil fuel usage would destroy the economy. In addition, energy efficiency was attacked as if it were communism by conservative think tanks, and Congress persistently blocked efforts made by the Clinton administration to extend requirements for SUVs' fuel efficiency. In the past decade, oil consumption has subsequently risen by 14 percent. According to Pope, by electing its own candidate, George W. Bush, as president, the oil-patch convention that dominates Congress aims to ensure an energy policy that is controlled by the oil industry.

Power to the People. Peter Fairly. *Technology Review* v. 104 pp70–77 May 2001

Fairly argues that fuel cells and microturbines could make everyone a power producer, relieving blackouts, lowering prices, and bringing electricity to the powerless. As people's dependence on electric and electronic systems grows, many businesses and consumers need better performance than the 99.9 percent reliability the local electric power grid provides. Deregulation frees consumers and unleashes a wave of investment and innovation in power technologies. The first products of this are clean, quiet, and dependable microturbines, developed in the 1960s to provide electric power for air-conditioning and circulation systems on aircraft. At least six kinds of fuel cells are also under development for electric power generation. The best hope for smaller, cheaper units lies in a light, compact version based on a structure called a proton exchange membrane; Ford, DaimlerChrysler, and Toyota are investing billions to make this type of fuel cell powerful and cheap enough to replace the internal combustion engine.

A Nuclear Nightmare. Douglas Pasternak. *U.S. News & World Report* v. 131 pp44–6 September 17, 2001

Pasternak explains how some U.S. nuclear plants are easy marks for terrorists. In the past decade almost half of the country's 103 power plants have failed mock terrorist attacks against them. In the past year alone, Nuclear Regulatory Commission inspectors have found alarms and video surveillance cameras that do not work, guards who cannot operate their weapons, and guns that do not shoot. Classified reports from Sandia National Laboratories demonstrate that a well-placed truck bomb would not even have to go into a site's property to destroy vital equipment, leading to a possible radiation release. Moreover, experts believe the water-intake systems at some plants are especially vulnerable to sabotage by either cutting off the water supply by clogging the intake valve or putting volatile chemicals into the reactor's cooling system.

Pumped for More Drilling: G. W. Bush. David Whitman. *U.S. News & World Report* v. 130 pp36–8 February 12, 2001

Republican senators, following up on President George W. Bush's campaign promise, hope to introduce a far-reaching energy plan in the near future. The 259-page draft of the Republican package contains several references to renewable energy sources and would increase spending on "clean-coal" technology and nuclear power sources. Bush and his colleagues are seeking to increase energy production from federal lands and speed up the approval process for oil and gas pipelines. The plan also controversially proposes opening up the coastal plain of the Arctic National Wildlife Refuge to oil drilling. During President Clinton's administration, more oil, natural gas, and coal were produced from federal offshore and onshore properties than during the administrations of Ronald Reagan or George Bush Sr. Challenges facing Bush's plans to increase energy production and cut dependence on foreign oil are discussed.

China Needs New Energy Technologies. Jeff Logan. *USA Today* v. 127 pp13–14 December 1998

Logan writes that China can afford to meet its growing energy requirements without compromising environmental aims. This can be achieved by using new technologies and accelerating market reforms, according to "China's Electric Power Options: An Analysis of Economic and Environmental Costs," jointly produced by Battelle's Advanced International Studies Unit in Columbus, Ohio, the Beijing Energy Efficiency Center, and China's Energy Research Institute. Lead author Jeff Logan says Chinese planners and government officials have regularly placed economic growth above environmental concerns. When considering the full environmental costs of production, it is actually more economical to use cleaner alternatives, such as flue gas desulfurization equipment, natural gas, and clean-coal technologies, he says.

A National Energy Plan: Diversity, Conservation, Harnessing New Technology. George W. Bush. *Vital Speeches of the Day* v. 67 pp482–5 June 1, 2001

President Bush addresses Capital City Partnership, St. Paul, Minnesota, on May 17, 2001. He asserts that, if Americans make the right choices now, they will meet their future energy needs in efficient, clean, convenient, and affordable ways. If they fail to act, the country could face a darker future, as reflected in increasing gas prices and rolling blackouts in California. Failure to act would also make the United States more dependent on foreign crude oil, giving foreign nations who may not share U.S. interests control of national energy security. The environment would also suffer, because government officials would strive to prevent blackouts by relying on more polluting emergency backup generators and by running less efficient, old power plants excessively. The president argues that America needs an energy plan that meets its energy challenges by protecting the environment, supplying increasing energy needs, and improving quality of life.

Index